Selling Ethnic Neighborhoods

Routledge Advances in Geography

1 **The Other Global City**
Edited by Shail Mayaram

2 **Branding Cities**
Cosmopolitanism, Parochialism, and Social Change
Edited by Stephanie Hemelryk Donald, Eleonore Kofman and Catherine Kevin

3 **Transforming Urban Waterfronts**
Fixity and Flow
Edited by Gene Desfor, Jennefer Laidley, Dirk Schubert and Quentin Stevens

4 **Spatial Regulation in New York City**
From Urban Renewal to Zero Tolerance
Themis Chronopoulos

5 **Selling Ethnic Neighborhoods**
The Rise of Neighborhoods as Places of Leisure and Consumption
Edited by Volkan Aytar and Jan Rath

Selling Ethnic Neighborhoods

The Rise of Neighborhoods as Places of Leisure and Consumption

Edited by Volkan Aytar and Jan Rath

NEW YORK LONDON

First published 2012
by Routledge
711 Third Avenue, New York, NY 10017

Simultaneously published in the UK
by Routledge
2 Park Square, Milton Park, Abingdon, Oxon OX14 4RN

Routledge is an imprint of the Taylor & Francis Group, an informa business

Typeset in Sabon by IBT Global.

First issued in paperback 2013

Library of Congress Cataloging-in-Publication Data

Selling ethnic neighborhoods : the rise of neighborhoods as places of leisure and consumption / edited by Volkan Aytar and Jan Rath.
p. cm. — (Routledge advances in geography ; 5)
Includes bibliographical references and index.
1. Ethnic neighborhoods. 2. Minority business enterprises. 3. Consumption (Economics) 4. Culture diffusion. I. Aytar, Volkan. II. Rath, Jan, 1956–
HT215.S45 2012
307.3'362089—dc23
2011025888

ISBN13: 978-0-415-71968-1 (pbk)
ISBN13: 978-0-415-89959-8 (hbk)
ISBN13: 978-0-203-17286-5 (ebk)

Contents

Figures

Introduction

Ethnic Neighbourhoods as Places of Leisure and Consumption

Volkan Aytar and Jan Rath

INTRODUCTION

Immigrant ethnic neighbourhoods—certainly those in Europe—are often associated with unemployment, vice and crime. With economic activity depressed and prospects uncertain, they easily become poverty traps. The concentration of 'urban outcasts' in the French 'banlieues', Dutch 'concentration neighbourhoods' and Danish 'ghettos' arouse concern, even suspicion (Wacquant 2008). But while representations of urban deprivation and decay speak their truths, they tell far from the whole story. For instance, recent years have witnessed the rise of ethnic neighbourhoods—notably their shopping areas—as sites for tourism, leisure and consumption in cities around the world. Brick Lane ('Banglatown') in London, Kreuzberg ('Klein Istanbul') in Berlin, la rue du Faubourg Saint Denis ('Quartier Indien') in Paris, and the Zeedijk ('Chinatown') in Amsterdam are only a few examples of ethnic neighbourhoods that attract tens of thousands of visitors from all corners of the world.

Ethnic entrepreneurs have played a key role in this development. Their social problems notwithstanding, these neighbourhoods have become breeding grounds for various forms of ethnic entrepreneurship, notably forms that are driven by the commodification of ethno-cultural diversity. Canoli, lamb vindaloo and halal meat, kitchenware and spices, flights and holidays, books and CDs, wedding dresses, massage oil and Buddha statues are all among the products that are—rightly or wrongly—associated with the cultural background of the entrepreneur. The commodification of ethno-cultural diversity constitutes these markets, creating new opportunities in otherwise blighted neighbourhoods.

The traditional countries of settler immigration have long acknowledged the economic potential of cultural diversity—witness the Chinatowns, Koreatowns, Greektowns, Manillatowns, Little Italys, Little Turkeys, Little Odessas and Little Saigons in cities like New York, San Francisco, Toronto, Vancouver, Perth and Sydney (Rath 2005). Once the run-down neighbourhoods of the marginalised, they now flaunt their ethnic diversity and are colourfully described in visitors' guides and on web sites to promote tourism and investment. The commodification of cultural diversity in these neighbourhoods not only attracts visitors, but also fuels employment, enhances livability, generates urban socio-economic development and fosters the branding of cities (Conforti 1996; Hackworth and Rekers 2005; Shaw, Bagwell and Karmowska 2004; Pang and Rath 2007; Rath 2007).

The growing tourism and leisure industries in these neighbourhoods offer opportunities to natives and immigrants, skilled and unskilled and males and females alike. They participate as organisers of cultural events, as web designers, as owners of cafés, coffee shops, restaurants, travel bureaus, hotels, souvenir shops, telephone and Internet shops, but also as waiters, cooks, dishwashers and janitors. These entrepreneurs and workers are the formal and informal caterers of entertainment and leisure. Together, they engender 'globalization from below' and, as Henry, McEwan and Pollard (2000) put it, 'create mainstream but unique products in terms of innovation, production and consumption'. The leisure economy thus constitutes an interface for immigrants from all social strata with the wider economy. In our globalising world—where local difference and place identity are increasingly important—heritage and cultural diversity have become crucial components of the cultural capital of post-industrial societies.

What accounts for this—perhaps unexpected—market-driven development? It may be tempting to explain the phenomenon in ethno-cultural terms: the rise of ethnic neighbourhoods as places of leisure and consumption is due to ethno-culturally specific entrepreneurial drives and the mobilisation of ethnic minorities' social and cultural resources (Ma Mung 1998; Wong 1998; Zhou 1992; see also Lin 1998 for a different perspective). But while such approaches are prominent within migration and ethnic studies, they fail to account for the emergence of specific economic opportunities. It should be noted, for instance, that the rise of ethnic entrepreneurship—certainly in Europe—coincides with the shift from industrialism to post-industrialism and the proliferation of economies that emphasise symbols, creativity and cultural consumption, while national and local authorities have increasingly adopted the re-imaging of (ethnic) neighbourhoods as strategies to revitalise urban districts and local economies. This is no accident. The implication is that any examination of these phenomena—any attempt to theoretically interpret and explain them—should go beyond studying the use of social, cultural and other resources, and embed this usage within wider economic and political contexts.

Mindful of the myriad connections between these transformations, this book explores the rise of ethnic neighbourhoods as centres of ethno-cultural commerce. More specifically, it explores manifestations of commodified cultural diversity by immigrant minorities in gateway cities and asks how these commercial expressions of culture have become vehicles for socio-economic development. In examining the role of ethnic entrepreneurs and workers in the leisure industry and their interactions with other actors, we see the emergence of a complex and dynamic ecological system that has dramatically affected urban political economies. We ask how these neighbourhoods have developed as sites of leisure and tourism, and examine the roles played by producers (entrepreneurs and their associations in particular), consumers, the structures that mediate cultural production and consumption, as well as local and national governments. We want to understand how these parties interact—and the intended, unintended and perverse effects for entrepreneurs, their respective ethnic groups, their neighbourhoods and cities at large—as well as how these more general processes play out in particular local contexts.

The involvement of immigrants in the urban tourist economy has received scant attention in the international literature. More than 20 years ago, Ashworth (1989) complained that despite its economic and social significance, urban tourism (in general) has been neglected by academic research. The situation has since changed (see for instance Hoffman, Fainstein and Judd 2003; Judd and Fainstein 1999; Hall and Williams 2002; Selby 2004). But while numerous studies have focused on the spatial, geographical, mental and symbolic meanings of tourism, leisure and consumption, they have not directly examined immigrant entrepreneurship (Crouch 1999; Löfgren 1999; Hall and Page 1999). Likewise, the literature on the economic integration of immigrant minorities in general and entrepreneurship in particular rarely, if ever, focus on tourism and the leisure industry. Most studies of Chinatown, Little Italy and other ethnic enclaves focus on their internal dynamics, and only mention in passing their potential for tourism, leisure or, more generally, their impact on the urban economy or the livability of cities (Laguerre 2000; Zhou 1992; see Kaplan and Li 2006 for a different perspective). Most studies moreover focus on single neighbourhoods so that local and national specificities are only casually addressed. International comparative studies are obviously few and far between.

More recently, several studies have connected immigration and migration to tourism, leisure and consumption (Hall and Williams 2002; Rath 2007), examining the impact on service sector employment in general and on tourism and leisure employment in particular, as well as the various 'human capital requirements' of these industries as they relate to demographic and urban change (Adler and Adler 2004; Aytar 2007; Iredale 2001; Li, Finlay and Jones 1998; Aitken and Hall 2000; Ram, Abbas, Sanghera and Hillin 2000; Rath 2005). Nevertheless, we cannot claim that these contributions have been enough to comprehend the diverse and complex interconnections between migration, immigration and ethnic entrepreneurship—the transformation of ethnic neighbourhoods into sites of leisure and consumption.

URBAN ECONOMIC TRANSFORMATIONS

We situate the rise of ethnic neighbourhoods as places of leisure and consumption and the role of ethnic entrepreneurs in the emerging tourism and leisure industry at the edge of different theoretical debates, roughly falling under the headings of urban political economy and the politics of culture. However, as the contributions to this book suggest, the distinction between urban political economy and the politics of culture is anything but clear. Indeed, one of our main concerns rests on the very fuzziness of the separation between theoretical discussions on political economy and the politics of culture. Let's unpack these theoretical debates.

We begin by focusing on the web of complex relationships centring on *ethnic entrepreneurs* and their position in the wider urban economy. As Kloosterman and Rath (2003; see also Kloosterman, Van der Leun and Rath 1999; Rath and Kloosterman 2000; Hall and Rath 2007) have argued while pursuing their 'mixed embeddedness' logic, ethnic entrepreneurship cannot be understood by solely looking at the social and cultural inclinations of immigrant populations. While factors such as entrepreneurial behaviour and competitive ethos obviously matter, ethnic entrepreneurship depends on a myriad of other factors, including changing market structures and the dynamics of supply and demand, urban renewal and revitalisation, the role of regulatory structures and larger governmental frameworks and the impact of the 'third sector' or actors within civil society.

Developments at the macro level provide the basic parameters of the fundamental changes that have taken place since the 1970s. Market structures and the dynamics of supply and demand have changed profoundly due to the transformation towards post-industrialism since the 1970s, and the implementation of neoliberal policies since the 1980s. Regulation School theorists see the 'regime of accumulation' and the associated 'mode of social and political regulation' between the 1930s and the 1970s as a 'Fordist-Keynesian consensus', based on mass production/consumption sustained by welfare states of various stripes, which in turn were supported by middle class values and lifestyles (for discussion see Lipietz 1986; Aglietta 1979; Boyer 1986). While this analysis has been challenged (see Amin 1994), its strength in explaining the post-1970s transformation of welfare states and the shift from Fordist towards 'flexible' forms of consumption should not be underestimated. Flexible production was coupled with the advent of the new knowledge, information and communication sectors and the shift from manufacturing to service sector employment.

These macro-level developments mattered for immigrant populations and their entrepreneurs. In the decades following the Second World War, large numbers of low-skilled and unskilled migrants were recruited under guest-worker schemes and allocated to 3-D positions in industries such as steel, meatpacking, automobiles, garments, shoes, shipbuilding and tobacco. Their employment enabled 'sunset industries' in the West to temporarily sustain their internationally competitive positions. In the 1970s, and especially after the first oil crisis in 1973, many of these industries were forced to shut down or relocate production to low-wage countries. Many workers were laid off and tottered into oblivion. Participation in the emerging post-industrial service economy, however, required educational qualifications and cultural competencies which immigrants often lacked, leaving them at the end of the employment line. Blocked labour market mobility spurred self-employment—one of the few alternatives to escape poverty and economic marginality (see for instance Waldinger *et al.* 1990). Ethnic entrepreneurship rose in many countries, more significantly as the post-industrial boom took off (Kloosterman and Rath 2003).

The rise of small-scale, flexible production and retail systems as well as the advent of service sector employment created an important window of opportunity for ethnic entrepreneurs (Sassen 1991). On the one hand, large corporations increasingly outsourced all sorts of tasks, including cleaning, catering, transportation, security, book-keeping, production, even research and development. On the other hand, high-income professionals enlarged the market for customised goods and services such as interior design, made-to-measure clothing, organic food and a plethora of other cool experiences. These two processes generated a complex putting-out system that spread both tasks and risks, involving small firms at both the higher and lower ends of the market. Ethnic entrepreneurs were heavily involved in these developments, albeit often at the lower end of the market (Kloosterman and Rath 2003).

The decline of manufacturing affected entrepreneurial opportunities in yet another way. Not every part of the production process was relocated to low-wage countries—the garment sector being a case in point (Raes 2000; Rath 2002). While the assembling of clothing was farmed out to offshore production facilities, design and marketing often stayed in the traditional fashion centres. Changing consumer preferences and marketing strategies, however, resulted in a breakdown of economies of scale. In an ever-changing, highly fragmented fashion market, retailers had to respond quickly to changes in fashion and unexpected increases in the demand for particular garments. This necessitated a production system capable of supplying small batches on short notice. In a number of fashion centres, local systems of subcontracting emerged involving a multitude of small contractors and home workers. Especially immigrants filled these slots. Lacking educational qualifications and capital, many ethnic entrepreneurs in this highly competitive market had to compete on price. Many were inclined to dodge taxes and cut corners in every possible way. Economic transformations thus also fuelled the informalisation of production (see also Kloosterman, van der Leun and Rath 1998).

Changes in production and consumption also altered patterns of their mediation. While functionalist mass consumption patterns gradually gave way to more 'pleasure-oriented', less utilitarian ones, reflexivity (as discussed by Giddens 1991, see also Lash and Urry 1994) became a major concern for many customers, with the aesthetisation of everyday life contributing to the reshaping of consumer tastes (see Featherstone 1991; Lash and Friedmann 1998). This led to the emergence and growth of distinct market segments with consumers seeking more symbolic fulfilment. Based on 'proto-communities of taste and style', lifestyle consumerism became the basis of status-based selfhood (Langman 1992). These trends as well as their spatial dynamics were highly relevant for the market opportunities of ethnic entrepreneurs. The residential concentration of immigrants fostered the spatial distribution of ethnic businesses and, at times, the creation of festivalised ethnic shopping areas and food centres. In the realm of urban symbolic economies, numerous scholars (including Zukin 1991, 1995, 1996 and 2009) have underlined the importance of the revalorisation of urban spaces—their re-invention as loci of leisure and consumption—a process in which neighbourhoods, districts and cities compete with one another. Especially in the context of the search for 'authentic' experiences among 'reflexive' consumers, the 'diversity dividend' of certain ethnic neighbourhoods have provided a 'competitive advantage'—as the contributions by Jordan and Collins, Novy, Pang, and Shaw and Bagwell suggest.

REGULATING THE ECONOMY

While the state's role as a 'comprehensive provider of services' (see Hall and Rath 2007) and its 'decommodifying', socially protective institutional layers may have weakened (Esping-Andersen 1990: 32), local and national government policies increasingly support place-promotion efforts. In the context of globalisation and the advent of 'global' or 'creative cities' (Sassen 1991; Florida 2000), public and private actors—often acting in concert—take great interest in urban symbolic economies (Ashworth and Voogd 1990; see also Novy's chapter on New York in this book). As the revalorisation of urban space is tied to the growing significance of spaces of consumption (Zukin 1995, 1996), the marketing of cultural diversity can become a key element in this process (Kloosterman and Rath 2003; Hall and Rath 2007).

We should also note that entrepreneurship in general, and ethnic or immigrant entrepreneurship in particular, does not develop in an institutional vacuum. It cannot be explained by the 'entrepreneurial proclivities' of certain groups. As Kloosterman and Rath (2003) and Engelen (2001) argue, the role of market forces and government regulation are always constitutive of the entrepreneurial environment. Understood as encompassing more than *legislation* or the *governmental* type of regulation (Engelen 2001: 210), regulatory mechanisms—zoning plans, tax measures, health and safety regulation, soft loan programmes, management support programs and so forth—can steer demand towards particular urban spaces, thereby supporting or hindering ethnic entrepreneurship and the development of ethnic neighbourhoods into sites of leisure and consumption. In Istanbul's Sulukule and Milan's Lazzaretto, official and public perceptions of ethnic populations played a negative role. In Istanbul, a self-sustaining entertainment economy was demolished by administrative fiat, while in Milan anti-immigration attitudes hindered the development of a leisure economy although other ingredients including 'diversity-seeking' consumers were present. In Perth's Northbridge, the local government was willing to promote the neighbourhood as a 'Chinatown' while many residents were not; similar dynamics were seen in London's 'Banglatown'. In other cases such as Martim Moniz in the neighbourhood of Mouraria in downtown Lisbon, the quotidian reality of immigrant populations was at odds with the government's 'imaging' of the city in general and this neighbourhood in particular. In New York's Harlem, the government's 'big-business friendly' tourism schemes created strains within the local population, some of whose members saw it as hindering community-led empowerment.

The impact of regulation (broadly understood) always exists in a dynamic relationship with the organised demands and initiatives of other societal groups. Here the role of the third sector—especially that of ethnic (entrepreneurial) associations and non-governmental and/or neighbourhood organisations—in empowering or otherwise mediating the dynamics of ethnic entrepreneurship deserves special attention. Numerous associations and alliances not only strive to place their own neighbourhoods on the urban political economy map; they also push regulatory bodies to become more accommodating of their constituencies' demands and expectations. The resistance to government and corporate-led efforts to ethnically brand neighbourhoods in London's Banglatown and Perth's Northbridge are interesting examples of the strength of the third sector, while NGO and community-based attempts to save ethnic entrepreneurship in Istanbul proved far less successful.

THE POLITICS OF CULTURE

The proliferation of ethnic neighbourhoods as places of leisure and consumption cannot be understood by focusing only on the political economy. Against the backdrop of what Vertovec and Wessendorf (2009) call the 'multicultural backlash'—the refusal of host societies to commit themselves to multiculturalism—public manifestations of ethno-cultural diversity cannot be taken for granted, even if they are economically promising. The contributions to this book therefore also address theoretical discussions around urban cultural politics.

Numerous theoretical debates not only underline the importance of the 'cultural turn' in the social sciences (Bonnell and Hunt 1999) but also the increasing social role of consumption as a symbolic and communicative mechanism (Miles 1998). Directly and indirectly conditioned by and in turn conditioning the above-mentioned changes in the political economy, mass consumption and its associated lifestyles gave way to individualised consumption patterns and niche markets catered to by the new service economy and its 'fun' sectors. The new market for cultural goods went hand in hand with the emphasis on the experience of consuming itself. Superimposed on the advent of consumer reflexivity, ethnic entrepreneurs aimed to capitalise on this development by refashioning their neighbourhoods as urban spaces providing 'authentic' experiences.

A number of gateway cities have responded to these developments by emphasising their multicultural character. Though such portrayals are often at odds with ongoing discussions on multicultural, integration or assimilation-oriented policies vis-à-vis immigrant and ethnic populations, one could argue that these cities are at least acknowledging multiculturalism—albeit in a rather utilitarian and sanitised form. They speculate on diversity as a resource for urban prosperity and as a catalyst for socio-economic development, particularly as investors consider diversity when locating their businesses. Other cities are wary or even thwart spatial-economic manifestations of ethno-cultural diversity; for them, even 'multiculturalism by stealth' is a no-go area. Shaped by the larger changes in culture and consumption, as well as by debates on citizenship, diversity and identity politics, the new urban politics of culture provides a fertile domain to discuss the role of ethnic entrepreneurs.

DIFFERENT CASES, DIFFERENT FACETS

We now turn to the empirical cases and theoretical analyses of the individual contributions. In his chapter on Harlem, Johannes Novy charts the history of ethnic urban tourism by comparing interest in the New York neighbourhood during the 1920s–1930s Harlem Renaissance and its recent rediscovery. Studying the urban ethnic tourism of an earlier era allows us to see what is *new* in the incorporation of communities in today's developments. Novy illustrates how New York's 'global city' ambitions have banked on post-industrial growth—'world class' business and producer services, real estate, culture, leisure and tourism, multiculturalism and diversity. The chapter also discusses the often controversial effects of regulatory frameworks and business-led initiatives on tourism in New York City, with critics pointing to the dangers of 'touristification' and supporters emphasising the opportunities for local communities. With rich examples, the paper shows how these developments have stirred debates around issues of race, ethnicity, class, community ownership and empowerment, cultural integrity and representation.

The following chapter by Stephen Shaw and Sue Bagwell examines the role of ethnic minority restaurateurs in revitalising Brick Lane ('Banglatown') in London's East End. Of the factors that helped transform a neighbourhood largely based on the 'rag trade' into 'London's Curry Capital', the authors stress the importance of narratives of place—promoted by alliances between local government, business and civil society shaped by the

New Labour government's 'Third Way' agenda. The chapter examines (i) the role of regulatory mechanisms, (ii) the partnership between local government, businesses and third sector organisations, and (iii) the tensions that arose in the community due to attempts to create streetscapes of consumption. It compares developments in Banglatown to two other multi-ethnic London thoroughfares—Green Street and Southall Broadway—and analyses their different strategies in terms of benefits for businesses, residents and community organisations.

In the following contribution, Ching Lin Pang examines the urban economies of 'Chinatowns' in Antwerp and Brussels. Her chapter relates urban economy to local government policy, as well as to the involvement of different types of ethnic entrepreneurs. Pang argues that although both ethnic precincts function as Asian ethnoscapes of leisure and consumption, there are significant differences in (i) the social and migration backgrounds of the Chinese entrepreneurs, (ii) their degree of embeddedness in local communities, and (iii) the respective neighbourhoods' positions within symbolic economies and broader urban renewal projects. The differences between the two cases should, Pang argues, alert us to the fact that ethnic precincts emerge and evolve along multiple trajectories in the post-industrial urban economy.

In a second contribution, Johannes Novy examines the role of urban ethnic tourism in the revalorisation of Berlin's Kreuzberg district. By addressing both the commodification of cultural and ethnic difference by immigrant populations and larger institutional/regulatory mechanisms, the chapter traces the development of Kreuzberg into an increasingly attractive site for tourism and leisure, which in turn affected how the neighbourhood is understood, represented, and treated by city officials. Nevertheless, the limited and sporadic initiatives to market Kreuzberg remain largely under-institutionalised—demonstrating the need for the efforts of ethnic entrepreneurs to be embedded within larger structures.

Volkan Aytar and Süheyla Kırca-Schroeder's chapter examines the informal rise and state-led demise of the local entertainment economy centred on the Roma neighbourhood of Sulukule, Istanbul. The chapter charts the development of Sulukule as an 'ethnicised' entertainment enclave in the Ottoman period, and how it became an informal alternative to 'learned' and 'well-behaved' forms of entertainment in the Republican era. It further shows how strategies of 'otherness' were instrumental in casting the Roma neighbourhood as an attraction both for Istanbulites and international tourists, and how in a parallel process of criminalisation, its self-sustaining entertainment economy was subject to police repression in the 1990s and neoliberal urban renewal projects in the 2000s. The chapter also examines the role of community organising and NGOs in resisting the latter plans, in promoting local entrepreneurial initiatives, and in internationalising the plight of the Roma, thereby addressing issues of discrimination, displacement and otherness within the politics of space.

Catarina Reis Oliveira's chapter focuses on Martim Moniz Square in the Mouraria neighbourhood of Lisbon, caught between the government's attempt to preserve it as a cultural heritage site and its contemporary use by Chinese entrepreneurs. Oliveira seeks to understand why contradictory discourses—often advanced by the same actors—developed in the course of several attempts to transform the district's image towards one of cultural diversity. In showing how immigrant entrepreneurial strategies to claim the space were constrained by considerations of heritage—especially the role of Martim Moniz and Mouraria as the spatial symbol of Lisbon's 're-conquest' from the Moors—the chapter adds a historical dimension to the politics of space within urban governance.

Kirrily Jordan and Jock Collins' chapter examines the strains caused by the attempts of entrepreneurs and city and state governments to market ethnicity in Northbridge, an inner-city district of Perth. While attempts to turn a section of Northbridge into a Chinatown faced strong local opposition, a more inclusive strategy now highlights the area's multi-ethnic character. Nevertheless, the authors show that the question of which groups are included in this re-visioning remains politically charged—the issues touch on the politics of representation, regulation and relations between and within ethnic groups.

In the final contribution, Roberta Marzorati and Fabio Quassoli discuss developments in the Lazzaretto neighbourhood of Milan, focusing on obstacles to immigrant incorporation in the urban economy and cultural politics. The authors argue that while appreciation of 'ethnic difference'—at least in terms of cuisine—prevails in Milan specifically and Italy more generally, this has not been complemented by changes in urban governance that would ease restrictions on, much less promote, ethnic entrepreneurial initiatives. Instead, the authors argue, a lack of strategic vision and anti-immigrant political and social discourses hamper the development of ethnic entrepreneurship.

TOWARDS NEW RESEARCH HORIZONS

This book aims to go beyond the presentation and analysis of current cases. In this concluding section, we summarise the basic findings of the contributions and point to new research horizons that may broaden the scope of studies on urban political economy and culture, as well as the study of ethnic neighbourhoods as places of leisure and consumption.

Group characteristics or 'entrepreneurial proclivities' cannot adequately explain the rise of ethnic or immigrant entrepreneurship, which is always embedded within larger social, economic, political and cultural structures (Kloosterman and Rath 2003; Kloosterman, van der Leun and Rath 1999). Focusing solely on 'ethnic' factors narrows the analysis to the micro level or, at best, to a single dimension of the meso level. Nor does it suffice to list more macro transformations, such as the shift to post-Fordist, flexible systems of production and consumption, or the rise of global and creative cities that create new entrepreneurial opportunities for ethnic and immigrant groups. As the contributions to this book suggest, we need to examine numerous—dynamic, interactive and contingent—factors to understand the rise or demise of ethnic/immigrant neighbourhoods in major cities.

Governmental and non-governmental actors help shape a complex arena of 'regulation' that promotes, allows, limits or blocks ethnic entrepreneurship. While certain macro transformations can expand opportunities for ethnic entrepreneurs, negative regulatory environments as well as societal and political biases against multiculturalism can hinder their development. In some cases, governmental actors may be willing to promote multiculturalism in general, or certain neighbourhoods as ethnic shopping/dining/entertainment areas in particular, but members of the ethnic group in question may resist such 'branding efforts', having already moved out of the 'lumpen bourgeoisie'. To better understand the many facets of ethnic entrepreneurship, our scholarly lenses need to take into account the varied roles of producers, consumers, the 'critical infrastructure' and (governmental as well as non-governmental) regulation. What does all this imply for future research?

First, while the rise of ethnic neighbourhoods may seem like a recent phenomenon, the social, economic and cultural 'functionalisation' of urban diversity and the links between this functionalisation and the leisure and consumption economies have histories spanning decades and, in some cases, centuries (see for instance Anderson 1990 and 1995; Lin 1998). However, this 'pre-history' of the role of ethno-cultural diversity in leisure and consumption has rarely been researched. Even less studied is the relationship between these ethnic groups, their neighbourhoods and the structures and processes of their adaptation to (or integration or assimilation into) mainstream societies and polities. These themes invite diachronic studies to complement existing synchronic ones, thus allowing comparison across cases as well as between various historical and more recent forms of ethnic neighbourhoods as places of leisure and consumption. Examples might include diachronic studies of various Chinatowns and other mono-ethnic neighbourhoods, of multi-ethnic neighbourhoods, and of the trajectories that lead to particular ethno-spatial constellations and economies.

Second, the political dimension of research on ethnic neighbourhoods as places of leisure and consumption, and more generally on ethnic entrepreneurship, remains weak (compare Rath 2000 and 2002; Engelen 2001; Kloosterman and Rath 2001). This is evident in at least two areas: (i) the role of regulation in its governmental and non-governmental forms; and (ii) the role of conflict and power among various actors. Regarding the former, regulation in the context of selling ethnic neighbourhoods is an under-researched area that could potentially contribute to theories of regulation in general. For the latter, further studies of power differentials among actors—including internal divisions and conflicts within ethnic groups and neighbourhoods—could bring deeper understanding of 'micro' dynamics and the relationships between the micro, meso and macro levels. Both could help to develop more effective policy for those working in the 'field'—be they local, regional or national government agencies, corporate actors; neighbourhood or community organisations; or the third sector in general.

Third, existing studies rarely venture a longer-term view on the viability of ethnic neighbourhoods, their communities, and their role in leisure and consumption. The 'rise' of such neighbourhoods, we have seen, is not a linear trend, and is always subject to disruptions. As numerous contributions to this book show, negative political and cultural attitudes threaten the integrity of many neighbourhoods, while others 'under-perform' even when the other ingredients for a successful outcome are present. The particular shape of this general 'rise' thus takes very different forms and may produce very different results. Future research would benefit from a longer-term perspective that takes into account contingencies and nuances as well as continuities and similarities.

This will allow scholars to produce more policy-relevant research on, for example, structural inequalities or neighbourhood-based development policies that can benefit immigrant groups. Studies linking governmental or non-governmental 'city branding' efforts to ethnic entrepreneurship would be particularly valuable, since in much of the writing on 'global' or 'creative cities', the creative industries, ethnic branding and urban tourism are seen as *Deus ex machina*. However, if there is a lesson to be derived from the contributions to this book, it is that such universally optimistic models can flounder on the rocks of local realities as well as systemic competition. Through interdisciplinary and comparative studies, we aim to achieve a much more thorough understanding of the structural dynamics of immigrant commercial activities, as well as the impact of immigrants on socio-economic developments more generally.

ACKNOWLEDGMENTS

This chapter is based on a series of individual and joint research endeavours. First, the research project on Ethnic Neighborhoods as Places of Leisure and Consumption carried out by Jan Rath with Annemarie Bodaar, Ilse van Liempt and Lex Veldboer at the Institute for Migration and Ethnic Studies (IMES) of the University of Amsterdam, sponsored by the Dutch Research Council NWO under the STIP program (nr. 473–04–317). Second, the project on Entertainment and Leisurely Consumption in Istanbul: Transformation of Spaces and Employment carried out by Volkan Aytar, initially involved with the State University of New York at Binghamton, then with the Turkish Economic and Social Studies Foundation (TESEV) and Bahçeşehir University in Istanbul, and later with IMES. Third, two international workshops (held in Rabat, Morocco, 10–12 May 2007 and

Istanbul, Turkey, 31 January–2 February 2008) were co-sponsored by NWO STIP, IMISCOE B6, NIMAR in Rabat, TESEV, the Dutch Consulate in Istanbul, and the Istanbul Journal of Urban Culture. All contributions to this book were initially presented at these workshops. The editors of the book and the authors of this introductory paper wish to express their gratitude to the sponsors and to the workshop participants.

REFERENCES

Adler, P.A. and Adler, P. (2004) *Paradise Laborers: Hotel Work in the Global Economy*. Ithaca and London: ILR Press.

Aglietta, M. (1979) *A Theory of Capitalist Regulation. The US Experience*. London: Verso.

Aitken, C. and Hall, C.M. (2000) 'Migrant and foreign skills and their relevance to the tourism industry', *Tourism Geographies*, 2(1): 66–86.

Amin, A. (ed.) (1994) *Post-Fordism: A Reader*. Oxford and Cambridge: Blackwell Publishers.

Anderson, K.J. (1995) *Vancouver's Chinatown: Racial discourse in Canada, 1875–1980*. Montreal: McGill–Queens University Press.

———. (1990) 'Chinatown re-roriented: a critical analysis of recent redevelopment schemes in a Melbourne and Sydney enclave', *Australian Geographical Studies*, 28: 131–54.

———. (1988) 'Cultural hegemony and the race-definition process in Chinatown, Vancouver: 1880–1980', *Environment and Planning D: Society and Space*, 6: 127–49.

Ashworth, G.J. (1989) 'Urban tourism: An imbalance in attention', in Cooper, C.P (ed.) *Progress in Tourism, Recreation and Hospitality Management*. London: Belhaven, 33–54

Ashworth, G.J. and Voogd, H. (1990). *Selling the City: Marketing Approaches in Public Sector Urban Planning*. London.

Aytar, V. (2007) 'Caterers of the consumed metropolis: Ethnicized tourism and entertainment labourscapes in Istanbul', in Rath, J. (ed.) *Tourism, Ethnic Diversity and the City*. (Contemporary Geographies of Leisure, Tourism and Mobility Series). London and New York: Routledge, 89–106

Bonnell, V.E. and Hunt, L. (1999) *Beyond the Cultural Turn: New Directions in the Study of Society and Culture*. Berkeley: University of California Press.

Boyer, R. (1986) *La Théorie de la Regulation. Une Analyse Critique*. Paris: La Decouverte.

Castells, M. (1989) *The Informational City: Information Technology, Economic Restructuring and the Urban-Regional Process*. Oxford: Blackwell.

Conforti, J.M. (1996) 'Ghettos as tourism attractions', *Annals of Tourism Research* 23(4): 830–42.

Crouch, D. (ed.) (1999) *Leisure/Tourism Geographies: Practices and Geographical Knowledge*. London and New York: Routledge.

Engelen, E. (2001) 'Breaking in and breaking out: A Weberian approach to entrepreneurial opportunities', *Journal of Ethnic and Migration Studies*, 27(2): 203–23.

Esping-Andersen, G. (1990) *The Three Worlds of Welfare Capitalism*. Cambridge: Polity Press.

Featherstone, M. (1991) *Consumer Culture and Postmodernism*. London: Sage Publications.

Florida, R. (2000) *The Rise of the Creative Class: And How it's Transforming Work, Leisure, Community, & Everyday life*. New York: Basic Books.

Giddens, A. (1991) *Modernity and Self-Identity: Self and Society in the Late Modern Age*. Stanford: Stanford University Press.

Hackworth, J. and Rekers, J. (2005) 'Ethnic packaging and gentrification: The case of four neighborhoods in Toronto', *Urban Affairs Review*, 41(2): 211–36.

Hall, C.M. and Page, S.J. (1999) *The Geography of Tourism and Recreation: Environment, Place and Space*. London and New York: Routledge.

Hall, C. M. and Rath, J. (2007) 'Tourism, migration and place advantage in the global cultural economy', in Rath, J. (ed.) *Tourism, Ethnic Diversity and the City*. (Contemporary Geographies of Leisure, Tourism and Mobility Series). London and New York: Routledge, 1–24.

Hall, C.M. and Williams, A.M. (2002) *Tourism and Migration: New Relationships between Production and Consumption*. Dordrecht: Kluwer Academic.

Harvey, D. (1989) *The Condition of Postmodernity*. Cambridge and Oxford.

Henry, N., McEwan, C. and Pollard, J.S. (2000) 'Globalization from below: Birmingham—postcolonial workshop of the world?', *Area*, 34(2): 117–27.

Hoffman, L.M., Fainstein, S.S. and Judd, D.R. (eds) (2003) *Cities and Visitors: Regulating People, Markets, and City Space*. Oxford: Blackwell.

Iredale, R. (2001) 'The migration of professionals: Theories and typologies,' *International Migration*, 39(5): 7–24.

Judd, D.R. and Fainstein, S.S. (eds) (1999) *The Tourist City*. New Haven: Yale University Press.

Kaplan, D.H. and Li, W. (eds) (2006) *Landscapes of the Ethnic Economy*. Lanham: Rowman & Littlefield.

Kloosterman, R. and Rath, J. (eds) (2003) *Immigrant Entrepreneurs: Venturing Abroad in the Age of Globalization*. Oxford and New York: Berg.

———. (2001) 'Immigrant entrepreneurs in advanced economies: Mixed embeddedness further explored', *Journal of Ethnic and Migration Studies*, Kloosterman, R. and Rath, J. (eds) *Special Issue on 'Immigrant Entrepreneurship*, 27(2): 189–202.

Kloosterman, R., van der Leun, J. and J. Rath (1999) 'Mixed embeddedness: (In) formal economic activities and immigrant businesses in the Netherlands', *International Journal of Urban and Regional Research*, 23(2): 253–67.

———. (1998) 'Across the Border: Economic Opportunities, Social Capital and Informal Businesses Activities of Immigrants', *Journal of Ethnic and Migration Studies*, 24: 239–58.

Laguerre, M.S. (2000) *The Global Ethnopolis: Chinatown, Japantown and Manilatown in American Society*. New York: St. Martin's Press.

Langman, L. (1992) 'Neon cages: Shopping for subjectivity,' in Shields, R. (ed.) *Lifestyle Shopping: The Subject of Consumption*. London and New York: Routledge, 40–82.

Lash, S. and Friedmann, J. (eds) (1998) *Modernity and Identity*. Oxford and Cambridge: Blackwell.

Lash, S. and Urry, J. (1994) Economies of Signs and Space. London, Thousand Oaks and New Delhi: Sage.

———. (1987) *The End of Organized Capitalism*. Cambridge: Polity.

Li, F.L.N, Finlay, A.M. and Jones, H. (1998) 'A cultural economy perspective on service sector migration in the global city: The case of Hong Kong,' *International Migration*, 36(2): 131–50.

Lin, J. (1998) *Reconstructing Chinatown: Ethnic Enclave, Global Change*. Minneapolis: University of Minnesota Press.

Lipietz, A (1986) 'New tendencies in the international division of labor: regimes of accumulation and modes of regulation,' in Scott, A. and Storper, M. (eds) *Production, Work, Territory: The Geographical Anatomy of Industrial Capitalism*. London: Allen and Unwin, 16–40.

Löfgren, O. (1999) *On Holiday: A History of Vacationing*. Berkeley: University of California Press.
Ma Mung, E. (1998) 'Territorialisation marchande et négociation des identités: Les Chinois à Paris', *Espaces et Sociétés*, 95: 65–81.
Miles, S. (1998) *Consumerism as a Way of Life*. London: Sage.
Pang, C.L. and Rath, J. (2007) 'The force of regulation in the Land of the Free: The persistence of Chinatown, Washington D.C. as a symbolic ethnic enclave', in M. Ruef and M. Lounsbury (eds) *The Sociology of Entrepreneurship* (Research in the Sociology of Organizations, Vol. 25). New York: Elsevier, 191–216.
Raes, S. (2000) *Migrating Enterprise and Migrant Entrepreneurship: How Fashion and Migration have Changed the Spatial Organisation of Clothing Supply to Consumers in The Netherlands*. Amsterdam: Het Spinhuis.
Ram, M., Abbas, T., Sanghera, B. and Hillin, G. (2000) '"Currying favour with the locals": Balti owners and business enclaves', *International Journal of Entrepreneurial Behaviour and Research*, 6(1): 41–55.
Rath, J. (ed.) (2007) *Tourism, Ethnic Diversity and the City*. London and New York: Routledge.
———. (2005) 'Feeding the festive city: Immigrant entrepreneurs and tourist industry', in E. Guild and J. van Selm (eds) *International Migration and Security: Opportunities and Challenges*. London and New York: Routledge, 238–53.
———. (ed.) (2002) *Unravelling the Rag Trade: Immigrant Entrepreneurship in Seven World Cities*. Oxford and New York: Berg.
———. (ed.) (2000) *Immigrant Business: The Economic, Political and Social Environment*. Basingstoke and New York: Macmillan/St Martin's Press.
Rath, J. and R. Kloosterman (2000) 'Outsiders' business: Research of immigrant entrepreneurship in the Netherlands', *International Migration Review*, 34(3): 656–80.
Sassen, S. (1991) *The Global City: New York, London, Tokyo*. Princeton: Princeton University Press.
Selby, M. (2004) *Understanding Urban Tourism: Image, Culture and Experience*. New York: I.B. Tauris.
Shaw, S., Bagwell, S. and Karmowska, J. (2004) 'Ethnoscapes as spectable: Reimaging multicultural districts as new destinations for leisure and tourism consumption', *Urban Studies*, 41(10): 1983–2000.
Vertovec, S. and Wessendorf, S. (eds) (2009) *The Multiculturalism Backlash: European Discourses, Policies and Practices*. London and New York: Routledge.
Wacquant, L. (2008) *Urban Outcasts: A Comparative Sociology of Advanced Marginality*. Cambridge: Polity Press.
Waldinger, R., Aldrich, H., Ward, R. and Associates (1990) *Ethnic Entrepreneurs: Immigrant Business in Industrial Societies*. Newbury Park: Sage .
Wong, B. (1998) *Ethnicity and Entrepreneurship: The New Chinese Immigrants in the San Francisco Bay Area*. Boston: Allyn and Bacon.
Zhou, M. (1992) *Chinatown: The Socioeconomic Potential of an Urban Enclave*. Philadelphia: Temple University Press.
Zukin, S. (2009) *Naked City: The Death and Life of Authentic Urban Places*. Oxford: Oxford University Press.
———. (1996) 'Space and symbols in an age of decline,' in King, A.D (ed.) *Re-Presenting Cities: Ethnicity, Capital and Culture in the 21st Century Metropolis*. New York and London: Macmillan, 43–59.
———. (1995) *The Cultures of Cities*. Oxford and Cambridge: Blackwell Publishers.
———. (1991) *Landscapes of Power: From Detroit to Disneyworld*. Berkeley: University of California Press.

1 Urban Ethnic Tourism in New York's Neighbourhoods

Then and Now

Johannes Novy

> There are two faces to New York. First of all it is a city of business, a gigantic anthill, throbbing with life. . . . But it is also a city of memories, a city where one can stroll, because the tourist with perceptive eyes discovers hidden treasures: urbanism, history, migration and so on. It seems that the different races of the world decided to meet here. From the Negro quarter of Harlem to the Ghetto, from the Italian and Polish slums to Chinatown, everything awakens tourism's curiosity. Each of these peoples' lives there transplanted but untransformed, carefully preserving the customs of their ancestors.
>
> *New-York. La Ville Merveilleuse. Revue de Voyages* (quoted in Gilbert and Hancock 2006: 96)

> [S]ome Harlemites thought the millennium had come. . . . I don't know what made any Negroes think that—except that they were mostly intellectuals doing the thinking. The ordinary Negroes hadn't heard of the Negro Renaissance. And if they had, it hadn't raised their wages any. As for all those white folks in the speakeasies and night clubs of Harlem—well, maybe a colored man could find some place to have a drink that the tourists hadn't yet discovered.
>
> Langston Hughes (1940: 228)

> The '20s are gone and lots of fine things in Harlem night life have disappeared like snow in the sun—since it became utterly commercial, planned for the downtown tourist trade, and therefore dull.
>
> Langston Hughes (1940: 227)

INTRODUCTION

Ethnic urban tourism is sometimes seen as a phenomenon of the late twentieth and early twenty-first centuries, a result of recent demographic, economic and political transformations that have fostered new patterns of leisure and consumption and changed the ethnic makeup of cities across the advanced capitalist world. But ethnic urban tourism is by no means a new phenomenon. In New York and Chicago, for instance, it was already fashionable in the mid-nineteenth century to visit segregated urban areas where the ethnic and foreign-born were concentrated—to go 'rubbernecking' (see Gates 1997; Cocks 2001; Gilbert and Hancock 2006).

New York's Chinatown became the object of a considerable slumming craze in the second half of the nineteenth century (Lin 1998b). Neighbourhoods such as the Jewish Lower East Side, Little Italy and the Black Tenderloin District followed suit a few decades later, while tourism in Harlem became a mass phenomenon in the 1920s and 30s when white (upper-) middle class New Yorkers and tourists crowded into Uptown jazz clubs and cabarets to explore the neighbourhood's nightlife and experience the creative spirit and excitement of the 'Harlem Renaissance'.

The popularity of neighbourhoods like Harlem and Chinatown as destinations between the mid-nineteenth and mid-twentieth centuries demonstrates, as Gilbert and Hancock (2006) point out, that racial and ethnic diversity was a key part of New York's appeal as the city joined Rome, London and Paris as one of the world's iconic urban tourist landscapes. 'As New York developed as a tourist city, "slumming"—sight-seeing trips into poor districts—became both more organised and standardised, and the spectacle of ethnicity and race became an important "sight" in the city, as significant in its own way as the Statue of Liberty or the Empire State Building' (Gilbert and Hancock 2006: 4).

New York has undergone profound changes since it became one of the world's great international tourist destinations. Yet the city's ethnic and racial diversity in the form of ethnic precincts, restaurants, shops and celebrations represent one of the city's most sought-after aspects. As Hughes' comments above illustrate, urban ethnic tourism remains a source of both great hopes and equally great disappointments.

Unlike in the past when tourism was viewed as little more than a marginal social activity, tourism today (along with leisure and consumption more generally) is considered a driver of contemporary urban change in New York, a critical source of revenue for the city and its neighbourhoods. As New York transformed into a multi-ethnic, post-industrial city centred on finance, real estate, culture and leisure in the second half of the twentieth century, tourism played an integral part in the city's transition. New York City's ethnic and minority neighbourhoods, which despite their appeal to earlier generations of visitors were considered anachronistic by the city's elites, came to be seen as vital economic and symbolic resources.

Embedded in a discussion of the political-economic, demographic and cultural changes that encompass urban ethnic tourism, the chapter first explores the origins and evolution of tourism and leisure in New York's ethnic and minority neighbourhoods from the beginnings of commercialised urban tourism up until the mid-twentieth century. It then examines ethnic and minority neighbourhoods' recent revalorisation as tourist destinations to discuss some of the key issues raised by the development of urban ethnic tourism in the advanced capitalist world. We focus on the neighbourhood of Harlem in northern Manhattan as its development exemplifies many of the different—and frequently contradictory—characteristics and effects associated with tourism in ethnic and minority neighbourhoods.[1]

TOURISM AND ETHNICITY IN NEW YORK: THEN

The rise of urban ethnic tourism is the product of profound changes in American cities and culture over the second half of the nineteenth century. Two developments deserve particular mention: (i) the urbanisation and industrialisation that transformed American cities such as New York from relatively small, homogenous towns into modern, sprawling metropoles with large immigrant populations; and (ii) the development of a commercial urban culture centred on leisure and consumption, together with the transformation of American cities into domestic and international tourist destinations (see Cocks 2001). These changes laid the groundwork for the emergence of New York as a tourist destination. Immigration from abroad and the northward migration of freed slaves from the southern states—alongside increasing segregation of the ethnic and foreign-born and the resulting pronunciation of cultural difference—set the stage for ethnic and racial diversity to become exploitable resources.

For politicians of the time, 'ghettos' such as Chinatown, Little Italy and the Lower East Side represented blighted, dangerous places standing in the way of modernisation and cultural assimilation. For many visitors, however, they were exotic and alien places where the foreign could be observed, dripping with exoticism and lurking danger. As tourism developed in these neighbourhoods, it steadily became more organised. Guidebooks and magazines began to feature New York's impoverished and 'colored' quarters as well as their perceived pleasures and horrors, while self-appointed guides began offering tours for 'slumming parties' (*New York Times* 1884). By the beginning of the twentieth century, so-called 'rubberneck automobiles' took spectators on journeys into New York's nether parts (Tracey 1905: SM 4). By the 1910s, about 200,000 'strangers' were visiting New York each day; in the words of the *New York Times* (1912: SM4), the city had become 'the Greatest Summer Resort in the United States'. Alongside Manhattan's iconic skyline and the other expressions of modernity that shaped New York's image as an archetypal twentieth-century city, the 'spectacle of poverty as well as of ethnic authenticity and exoticism' (Gilbert and Hancock 2006: 97) became a primary attraction for visitors 'doing the town'.

Members of New York's emerging leisure class meanwhile expressed shock and dismay over the class and ethnic stratification that had come to define their city, as well as over the crime and social unrest in its slums. That said, they were fascinated by the mix of despair, exoticism, pleasure and vice inherent in them. Many nineteenth-century descriptions reveal that explorations of New York's 'low life' (Sante 1991) were inspired by both voyeurism and a developing sense of the 'pleasures of cosmopolitan consumption' (Gilbert and Hancock 2006: 98). While xenophobia and racism intensified during the period and many New Yorkers were troubled over the growing number of foreign faces in their midst (Cocks 2001: 194), ethnic and racial differences were increasingly regarded as curiosities and commodities as a new cultural emphasis on leisure and consumption gradually took hold. The appeal of those neighbourhoods where the ethnic and foreign-born were concentrated was heightened by the image of insular 'villages' where 'primitive' inhabitants withstood the forces of modernity to offer visitors 'pure' and 'authentic' experiences in an otherwise relentlessly transforming metropolis (Dowling 2007).

It did not take long for those areas hailed as particularly 'authentic' and 'genuine' to become staged. Residents began using tourism for their own ends as New York's bourgeoning tourism and leisure economy moved in. Chinatown's development was a case in point: as early as the beginning of the twentieth century, tourists had begun complaining that the area was losing its appeal (Gilbert and Hancock 2006). The *New York Times* lamented that touring the nearby Bowery, once a paradise for slumming and rubbernecking, was 'very much like seeing Pompeii: One must rely on imagination and reading to get an idea of the real thing' (Tracey 1905: SM4). The residents of Chinatown meanwhile grew increasingly unnerved over tourism and the stereotyping and voyeurism it involved. One local businessman complained about the 'barkers' on sightseeing buses: 'They relate stories of crime that never took place. They characterize the homes of respectable Chinese . . . as opium joints. They point to any building at random and inform visitors that . . . murderer . . . hid there. We have had enough' (*New York Times* 1923: 16).

DESTINATION HARLEM: THEN

Harlem did not attract much attention as a tourist destination until the early 1900s when the immigration of blacks from other parts of New York, from the South, and from Africa and the Caribbean turned the neighbourhood into the 'capital of the African Diaspora' (Maurrasse 2006: 17), or in the words of James Weldon Johnson (1991: 2), the 'greatest Negro city in the world'. What had been a suburban community for the upper classes in the north of Manhattan became a 'neighbourhood transformed', 'housing 50,000 Blacks in 1914, and nearly 165,000 by 1930'. As Harlem developed, artistic and intellectual activity began to flourish in the soon legendary neighbourhood (Sacks 2006: 3).

This 'Harlem Renaissance', as the period between the 1920s and 1930s soon became known, 'not only fostered a sense of pride among people of African descent, it also caught the attention of Whites' (Maurrasse 2006: 21). By the mid-1920s, Harlem had acquired a 'world-wide reputation' (Johnson 1991: XIV): white Americans and foreign tourists flocked to the neighbourhood to visit jazz clubs and cabarets, experience the area's lively street culture, and consume alcohol as Prohibition laws were less effectively enforced in Harlem than in other parts of the city. A sizable tourism and leisure economy thus developed, employing thousands of residents and reinforcing Harlem's image as an entertainment destination. At the same time, Harlem's lure also obscured the hardships experienced by most of its residents. While Harlem's entertainment industry was indeed thriving, 'much of the ownership of the Harlem institutions of the time rested in white hands' (Maurrasse 2006: 22). Most residents saw little of the thousands of dollars spent night after night in the neighbourhood's cabarets and clubs. What had been created in Harlem was, according to Gilbert and Hancock (2006: 99), 'a many layered fantasy world of black life and culture' that effectively trivialised the neighbourhood's social and racial tensions. Over the years, more and more Harlemites grew concerned over the exploitive nature of the tourist trade in their midst. Their concerns were reflected in Langston Hughes' *The Big Sea*, which illustrated their awareness that the growing presence of white tourists was threatening the area's main 'selling point'—its 'blackness':

> ordinary Negroes [did not] like the growing influx of whites toward Harlem after sundown, flooding the little cabarets and bars, where formerly only colored people laughed and sang, and where now the strangers were given the best ringside tables to sit and stare at the Negro customers—like animals in a zoo. . . . Some of the owners of Harlem clubs, delighted at the flood of white patronage, made the grievous error of barring their own race. . . . But most quickly lost business and folded up, because they failed to realize that a large part of the Harlem attraction . . . lay in simply watching the colored customers amuse themselves (Hughes 1940: 221).

Resentments over Harlem becoming a white playground were fuelled by the unaltered discrimination and lack of opportunities confronting African Americans. The area's living conditions deteriorated significantly in the late 1920s and early 1930s when overcrowding reached new heights and the Great Depression set in, leaving half of Harlem's residents unemployed and two-fifths of the neighbourhood's families on government relief (Greene 1979: 514; Osofski 1963). As Harlem's fortunes declined, the neighbourhood's allure as a joyful place of 'laughing, singing and dancing' (Johnson 1991: XIV) faded; disinvestment, deterioration, poverty and social disorder took its place.

By the 1960s, over 25 per cent of Harlem's housing stock was dilapidated beyond rehabilitation while the neighbourhood was widely considered too dangerous to visit. Harlem had transformed into Marcuse's (1998) 'outcast ghetto' where the marginalised, unemployed and unwanted involuntarily congregated, while those who could afford to left the area. With factory jobs disappearing, urban renewal projects threatening the integrity of the built environment and residents fleeing the inner city by the thousands, many ethnic neighbourhoods in the city were experiencing a similar fate. Some vanished altogether as a result of slum clearance, assimilation and slowing immigration streams. By the end of the 1960s, analysts nationwide were speaking of an urban crisis as poverty, crime, racial tensions, economic distress and social pessimism intensified in cities across the country. New York's social, political and economic problems loomed particularly large; the city became a 'metaphor for all of the nation's intractable urban ills' (Sagalyn 2001: 5). The 'Fun City', as then Mayor John Lindsay cheerfully labelled New York (Gates 1997; Greenberg 2008), became known as 'Fear City'; visitor numbers dropped while urban ethnic tourism in particular came to be seen as a thing of the past. Many consumable aspects of ethnicity—previously confined to specific urban areas—had become part of the American mainstream; ethnic neighbourhoods that had previously attracted tourists lost their vitality or disappeared, while a racially charged culture of fear took hold of the city.

TOURISM AND ETHNICITY IN NEW YORK: NOW

Present-day New York bears little resemblance to the city during the doom and gloom of the 1970s and 1980s when it was written off as an 'obsolete anachronism of an earlier age' (Sagalyn 2001: 6). New York today is considered one of the world's few true 'global cities' (Sassen 1991) and a 'frontrunner' of post-industrial urban growth (Savitch and Kantor 2002: 6–7), fuelled by business and services, real estate, culture, leisure and tourism.

Immigration contributed to the city's economic recovery by replenishing its labour force, creating new jobs and industries, strengthening its transnational ties, and turning old, decaying neighbourhoods into 'polyglot honeypots' (Lin 1998a) that would play pivotal roles in the new urban economy.[2] Reflecting the widely noted revalorisation of multiculturalism and diversity—seen in the growing popularity of ethnic-themed food, music, literature and clothing—ethnic and minority neighbourhoods throughout the city (again) become popular destinations for tourists and New York's sizable 'cosmopolitan consuming class' (Fainstein et al. 2003: 243). Alongside the city's established ethnic attractions of Chinatown and Little Italy in downtown Manhattan, which never ceased to attract visitors (Lin 1998b; Conforti 1996), other neighbourhoods established themselves as parts of New York's leisure landscape. Harlem after decades of abandonment also witnessed more visitors.

While new immigration streams, the proliferation and rejuvenation of New York's ethnic urban spaces and increased consumer demand for ethnic diversity were key factors in the resurgence of urban ethnic tourism, on their own they cannot explain the resurgence of ethnic and minority

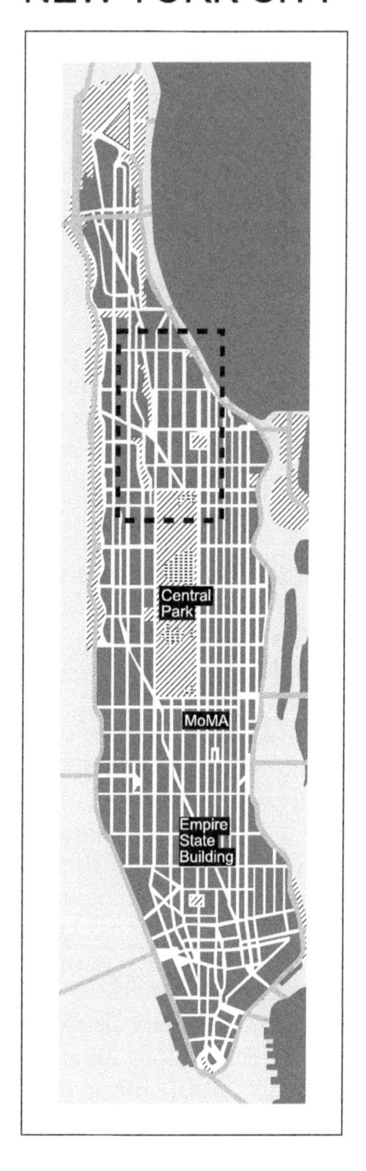

Figure 1.1 Location of Central Harlem in New York City. *Map designed by Ana Sala.*

HARLEM

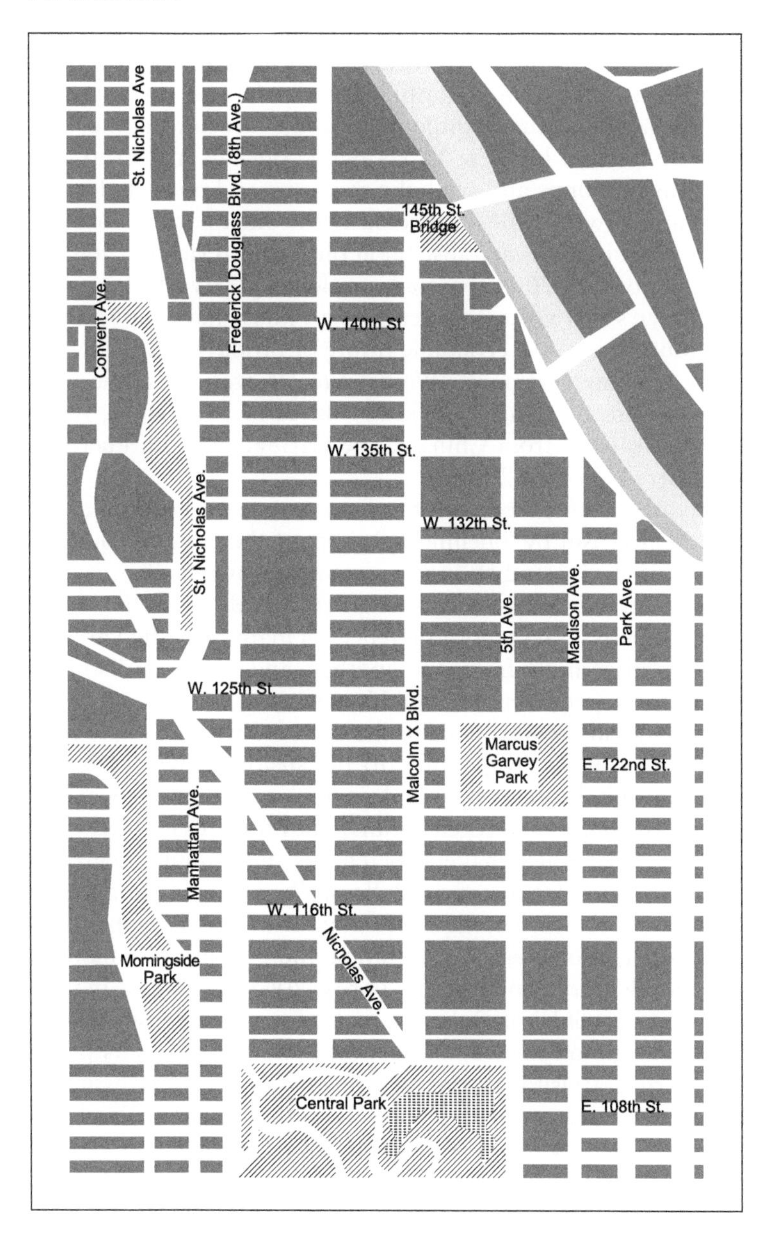

Figure 1.2 Map of Central Harlem. *Map designed by Ana Sala.*

neighbourhoods as sites for tourism and leisure in the late twentieth century. Rather, the phenomenon is entwined with a broad range of factors that have altered the development context of ethnic and minority neighbourhoods in cities around the world. Mirroring what has been observed elsewhere (Shaw *et al.* 2004; Rath 2005; Lin 1998a), and in contrast to the early twentieth century, ethnic spaces today are not shunned, excluded or otherwise ignored by New York's public officials and local elites, but recognised as economic and symbolic resources and promoted accordingly (Fainstein and Powers 2006). Particularly the offices of the borough presidents have attempted to capitalise upon the cultural richness found within their jurisdictions. In 2004, borough president Marty Markowitz inaugurated Brooklyn's first visitor centre. 'We all know how important attracting more tourists is to the economic future of Brooklyn', Markowitz stated during the opening ceremony. 'We need to let everybody know that you can see the entire world and stay in Brooklyn: in 30 minutes, you can get from Poland to Italy, from China to the Caribbean, from Puerto Rico to Pakistan' (Brooklyn USA 2004).

Equally important, decreasing crime and gentrification encouraged urban ethnic tourism by 'opening up' parts of the city previously deemed too dangerous to visit and by increasing outside awareness, (media) exposure and the attractiveness of many neighbourhoods. Gentrification further fuelled consumers' 'craving for difference' as many gentrified areas seemed less and less discernible from one another, and also less distinguishable from the suburbs (see Hammett and Hammett 2007).

Local policies also contributed to the rise of tourism in many ethnic and minority communities. Responding to the steady growth of New York's leisure and hospitality industry[3] and mirroring the more general shift of economic development and regeneration policy towards culture, leisure and consumption, local policy-makers, merchants and community groups made concerted efforts to capture tourist dollars. The Business Improvement Districts (BIDs), which arose in the 1980s and have since developed into influential drivers of commercial revitalisation and expansion in more than 50 neighbourhoods throughout the city (Gross 2005), are a case in point.[4]

The Lower East Side's BID, for example, aims to lure visitors by capitalising on the neighbourhood's multicultural past as well as its present-day diversity. It runs a visitor centre, organises walking tours, issues promotional brochures and has invested heavily in the neighbourhood's physical appearance. Other local actors promote tourism not for its own sake or monetary gain, but to showcase ethnic or minority communities' achievements, preserve (multi-)cultural heritage, nourish civic pride and bring people of different ethnicities, classes, lifestyles and backgrounds into contact with one another. Another good example of a neighbourhood-based effort is the Community Development Corporation 'The Point' which promotes tourism in the South Bronx to draw attention to the Dominican, Puerto Rican, Cuban and African American neighbourhood's rich musical history ('from Mambo to Hip Hop') and to challenge the South Bronx' stigmatisation as a 'no-go-area' (Fainstein and Powers 2006).[5]

The 'From Mambo to Hip Hop Tour' is only one of many recent attempts by South Bronx businessmen, politicians and community activists to spearhead the development of tourism. Borough President Carrion Jr. regularly emphasises its significance for the neighbourhood's future, while the Empowerment Zone (EZ)—the United States' only remaining federal initiative devoted to neighbourhood development—has invested in several small and mid-sized cultural tourism projects and related initiatives (Hostetter 2002). That said, the case of the South Bronx also reveals the obstacles faced by urban ethnic tourism and local efforts to benefit from the city's bourgeoning tourism and leisure economy (see Fainstein and Powers 2006).

The growth of ethnic tourism and the facilitation of leisure beyond the city's main tourist areas notwithstanding, a complex set of 'structural and contingent factors' places New York's ethnic and minority neighbourhoods 'at a structural disadvantage for attracting tourists' (Fainstein and Powers 2006: 150). Some have to do with New York's peculiar geography and transport system which limit the accessibility of the outer boroughs that contain most recent areas of ethnic concentration. Signage is insufficient; even New Yorkers find it difficult to orient themselves in unfamiliar areas outside Manhattan. Crossing the East River also creates 'a psychological and—to some extent physical barrier' (*Ibid.*). As most popular tourist sites and hotels are in Manhattan, particularly out-of-towners are unlikely to wander into other areas.

Other obstacles include ethnic and minority communities' historic neglect in urban development and governance, the city government's bias towards Manhattan, the favouring of large over small business interests and recent trends in redevelopment programmes that favour large-scale urban renewal at the expense of community-oriented planning (Fainstein 2005; Fainstein and Powers 2006). 'Instead of seeking to foster heterogeneous centers and to offer opportunities to immigrant entrepreneurs', recent redevelopment policy threatens many of the city's thriving communities (Fainstein 2007: 164). While some of the city's redevelopment projects are likely to lure visitors to previously neglected areas, such as the new Yankees Stadium in the Bronx, most have attracted considerable local opposition, involve primary and secondary displacement, and are deemed insensitive to local communities. The marketing efforts of the borough presidents' offices meanwhile often fall short of their rhetoric, and many local actors lack the resources and political clout to successfully market their communities or to challenge the Manhattan-centric view of the city. While NYC & Company—the city's official tourism marketing organisation—has recently tried to draw attention to the outer boroughs and their ethnic diversity (Fainstein and Powers 2007; Collins 1998), most such efforts have been small-scale; the overall marketing of the city remains, in the words of Miles and Miles

(2004: 52), 'as reductive as the skyline of a postcard view from a distant vantage point'.

Moreover, some local actors perceive the development of tourism and the leisure economy as detrimental to their communities' well-being. Tourism is often seen as culturally destructive and economically exploitative, intimately tied to gentrification and the eventual displacement of the poor. These and similar concerns have only been galvanised by the influx of affluent visiting New Yorkers, who give residents the—often correct—impression that they are scouting for new areas to gentrify (Gates 1997).

DESTINATION HARLEM: NOW

Struggling with blight and neglect, Harlem was until the 1980s considered a 'no-go area' by New Yorkers and visitors alike. In the late 1970s, the *Chicago Tribune* opined that Harlem was 'distinguishable by dilapidated buildings, burned-out tenements, and other unlovely sights, toured only by social workers . . . and by politicians at election time' (Oppenheim 1978: 48). While a local tour operator, the Penny Sightseeing Company, had inaugurated guided tours of Harlem in the late 1960s (MacCannell 1999: 40; Sandford 1987), such trips remained marginal. The area was simply considered too dangerous and remained off-limits to most visitors. The few existing offerings exemplified what MacCannell termed 'negative sightseeing', where the attraction was poverty and devastation.[6]

Harlem's image as a 'ghetto' was anathema to its business community. The first attempts to challenge the neighbourhood's negative image and replicate its earlier success as a tourist destination date back to the mid-1970s (Rule 1979: B3) when, sponsored by the Uptown Chamber of Commerce, the first local tourist maps were printed in more than 50 years. Of the projects designed to draw tourists, some were successful like the inauguration of the annual 'Harlem Week' (which first took place in 1974 as Harlem Day). Others failed, including efforts to create a local marketing organisation and visitor centre. However, it was not until the 1980s that observers began to see increasing tourism and leisure activity as particularly European and Asian visitors began venturing into Harlem in growing numbers to attend gospel and jazz concerts, search for traces of the 'Harlem Renaissance', experience the neighbourhood's vibrant street life, explore the areas' numerous historic sites and—undeniably a motive for some—satisfy their curiosity of the neighbourhood's alleged 'ghetto culture' (Severo 1984; Chira 1989). The 'epitome of all that is dangerous and hopeless about urban America', the *New York Times* proclaimed in 1989 (Chira 1989: A1), was becoming 'a must-see stop for tourists curious about black music, food, and culture'.

Fuelled by such boosterism and resurgent interest in Black America and the traditions (classic jazz, funk, etc.) and trends (hip hop, graffiti, etc.) it was associated with (Hoffman 2003), Harlem began to attract an increasingly diverse set of visitors in the late 1980s and 1990s, including a growing number of domestic tourists (Audience Research and Analysis 2000). This growth, as Hoffman (2003: 288) points out, 'preceded the development of a tourism infrastructure and the decline in the crime rate' and 'jump-started a fledgling tourism industry, giving rise to Harlem-based tours and activities initiated by local . . . entrepreneurs as well as city-wide tour bus-operators'. Services catering (though not exclusively) to tourists such as restaurants and bars proliferated, while established cultural institutions and landmarks (e.g. the Apollo Theater and Studio Museum) were renovated; a number of new attractions were developed or proposed (e.g. the Jazz Museum and Museum for African American Art).

That said, the recent growth of Harlem's tourism infrastructure cannot be attributed to increasing demand alone. Development was spurred by the Upper Manhattan EZ legislation introduced in Congress by Harlem Representative Charles Rangel in 1994. Soon after its implementation, the legislation became a driving force in Harlem's revalorisation as a destination (Maurrasse 2006; Hoffman 2003). Explicitly treating Harlem's culture and ethnicity as a means for job and investment creation,[7] it provided tax breaks and loans for large-scale projects like the Harlem USA shopping and entertainment complex and other retail and entertainment establishments, steered marketing campaigns to attract visitors, encouraged the start-up or expansion of restaurants and other services and created grant and loan funds to nourish Harlem's cultural scene. Although critics charge that the EZ spent much of its funding inefficiently, it played an important role in opening the area to visitors and tourism-related investments. It did so not only through raw dollars but by making Harlem appear safe to investors and visitors, by creating a climate in which openness to tourism was seen as a key to Harlem's cultural and economic empowerment and by integrating Harlem into New York's wider urban entrepreneurial regime. By 2000, a study commissioned by the EZ found that 800,000 people were visiting Harlem each year, generating approximately $164 million annually (Audience Research and Analysis 2000). Harlem's 'ghetto economy' was being rewoven into the wider economy.

HARLEM'S 'SECOND RENAISSANCE' AND ITS DISCONTENTS

After decades of economic abandonment, Harlem today is home to trendy bars, restaurants and boutiques. Chain stores such as Starbucks have moved in while several hotels—the first in decades—are about to be built. Fuelled by New York's booming real estate market, the average price of residential homes—especially the neighbourhoods' historic building stock—has skyrocketed. Given Harlem's state of disinvestment only a few decades ago, the neighbourhood's current era of commercial development and residential desirability is welcomed by many residents for bringing safety, services and jobs, transforming the neighbourhood's image and empowering the community's political and civic spheres (Hoffman 2003). At the same time, Harlem's 'second renaissance', as some are calling it, has become a source of conflict—particularly the neighbourhood's 'touristification' (Maurrasse 2006: 31; Hoffman 2003: 293–94). Reminiscent of concerns expressed during the Harlem Renaissance, critics of the current transformation advance two intertwined arguments:

1) Harlem's revalorisation has not benefited but rather placed additional burdens on most of its long-term residents as 'the historic black working-class community threatens to become Manhattan's newest yuppie mecca' (Kelley 2007: 65).
2) Harlem's transition threatens to destroy many of the features that made it attractive as a destination in the first place (Chinyelu 1999; Dávila 2004; Maurrasse 2006).

Concerns over gentrification and displacement loom large in present-day Harlem. While the extent of displacement continues to be debated, solid evidence suggests that Harlem's most vulnerable groups are increasingly falling victim to the neighbourhood's revalorisation or face difficulties 'staying put' (Newman and Wyly 2006). Partly due to the concentration of public housing in the area, Harlem continues to be among New York City's most impoverished neighbourhoods, with over a third of residents receiving some sort of public assistance and low-income residents experiencing marginal socio-economic gains (Maurrasse 2006). At the same time, the real estate market affects the local business community, which not only faces increasing costs but heightened competition from larger businesses as well as a changing customer base. Currently facing a controversial rezoning, 125th Street today is increasingly dominated by national and international retail chains while many long-standing small businesses in the neighbourhood have difficulties adjusting to the transition and are frequently forced to close (Maurrasse 2006: 68–70; Eligon 2007). This is at least partly due to the Upper Manhattan Empowerment Zone's approach to economic development, which has been criticised for devoting too much attention to attracting outside capital and neglecting the needs of local residents and businesses (Maurrasse 2006; Chinyelu 1999; Oakley and Tsao 2006). Reflecting the widespread fear in the community about Harlem's 'colonisation', Chinyelu (1999) compares the EZ's approach to the exploitation of the 'Third World', while Kelley (2007: 68) accuses the EZ of pursuing a 'somewhat Darwinian free-market approach to the matter—only the strong survive and that's the way it ought to be'. Along similar lines, Dávila takes particular issue with the marketing of Harlem's ethnicity and culture, arguing that the EZ distorts and banalises

Harlem's culture and ethnicity by treating them as mere 'profit-making devices' (Dávila 2004: 50).

Though complaints about the dilution of Harlem's cultural integrity are regularly heard, no survey has systematically examined residents' attitudes towards tourism or Harlem's transition more generally. Recent rumours about the opening of a Harlem-themed restaurant on 125th Street offering 'Zora Neale Hurston salads', 'Miles Davis omelets' and 'Denzel burgers' (Pogrebin 2005) has not helped to calm such fears. For many residents, it is Harlem's black identity that is at stake, a loss of Harlem's soul that will ultimately threaten the community's future development, including its potential as a tourist destination. 'What could be found in Harlem', Maurrasse (2006: 39) observes, 'could not be found anywhere. A Gap, an HMV, a Starbucks, an Old Navy, and other similar chains can be found in just about any American city or suburban strip'.[8]

CONCLUSION

As I wrap up this chapter, two more neighbourhood institutions are being washed away by Harlem's transformation (Williams 2008). Bobby's Happy House, a record store opened by music producer Barry Robinson in 1946 on 125th Street, closed in January 2008 due to rent increases Robinson could not afford. Robinson, now in his 90s, was the first African American to ever own a store on 125th Street. A few hundred yards northwest, the Harlem Record Shack has been given until the end of March to vacate its premises of 36 years, to make way for a new retail and office building. Sikhulu Shange, the owner of the Harlem Record Shack, was not invited to set up shop in the new building. Both stores are renowned not only for their African, gospel, funk, hip hop, jazz and R&B selections but also, as Eligon (2007) puts it, as landmarks of Harlem's black heritage and places that residents could call home. Their eviction exemplifies a process only too familiar to scholars of tourism: the community is falling victim to its own success. Both stores had attracted individual tourists and guided tours for decades. 'With that being gone', the owner of Hush Tours told the *New York Times* (Williams 2008), 'it's really, really going to hurt the community and the preservation of the roots of hip-hop'.

A few blocks east, preparations have begun to transform the long-vacant Victoria Theater—once Harlem's largest and most elegant theatre—into a cultural centre with two live theatres, a jazz museum, a hotel and condominiums. Thus within a few city blocks, we witness the dynamics through which tourism can transform neighbourhoods. The developments have fuelled acute discussions surrounding issues of race, class, community ownership, cultural integrity, and (dis-)empowerment and representation, and exemplify the different—and sometimes contradictory—effects of tourism in ethnic and minority neighbourhoods. While some see Harlem's current transition as a 'massive displacement, relocation, ethnic cleansing program' (Chinyelu 1999: 125) threatening its identity as a black community, it seems more accurate to see the current transition as one that is as much about class as it is about race. Despite the influx of other ethnic groups, Harlem remains overwhelmingly African American; many of the protagonists—and beneficiaries—of Harlem's transition belong to Harlem's black elite. Urban ethnic neighbourhoods, the case of Harlem makes clear, cannot be conceived as monolithic entities. They consist of different groups with different backgrounds, needs and interests, affected differently by processes of neighbourhood change.

Urban ethnic tourism, this chapter has shown, has long been an important element within New York's tourism and leisure trade. The appeal of ethnic and minority neighbourhoods as destinations as well as the accompanying transformation of culture and ethnicity into 'marketable commodities' is hardly new. What is new is the intensity with which these previously marginalised communities are being incorporated into post-industrial urban economies and development cycles. As Hoffman (2003: 297) argues, Harlem's current transformation illustrates 'a new mode of regulation, making for greater social/political and economic inclusion, but with the associated costs as well as benefits'—costs which many consider too high as tourism frequently operates in contexts of urban development that disadvantage local communities' (particularly the most vulnerable within them) and their ability to control the negative effects. As this chapter has argued, tourism and leisure today—as in the late nineteenth and early twentieth century—is better understood in light of the broader institutional and regulatory frameworks in which communities find themselves, as well as the underlying power relationships by which they are structured. The challenge is to find ways to transform the framework—to expand the opportunities the tourist and leisure economies promise for ethnic minority neighbourhoods and the social groups within them.

NOTES

1. In the following discussion Harlem refers primarily to the historic African American community today known as Central Harlem. Occasional reference will be made to developments in West and East Harlem since the boundaries between these three districts are anything but rigid.
2. New York in recent decades has become a 'majority minority city'. The 2005 Community Survey Data of the US Census revealed that ethnic minorities make up over half of New York City's population. The city added nearly 800,000 residents from abroad in the 1990s alone. Without this gain in immigrants, the city's population would have decreased. Instead, it increased to 8,213,839 (up from 7.3 million in 1990). Nearly half of the city's foreign-born come from the Caribbean and Latin America, while significant numbers also hail from Eastern Europe and East Asia. Over the same period, the percentage of foreign-born New Yorkers rose from 28.4 to 35.9 per cent; 52.6 per cent are from Latin America while 23.9 per cent are from Asia.

3. In 2005, the city's leisure and hospitality sector (including the arts, entertainment, food and accommodations) was estimated to support 333,128 of the city's approximately 3.6 million workers, 85,000 more than at the beginning of the 1990s (Office of the New York Comptroller 2006). An estimated 135,000 jobs directly depended on tourism, while tourism-related employment across sectors was 283,500—an annual growth of 2 per cent over the last 15 years. According to the report, spending on accommodation, food and entertainment totalled some $10.1 billion. Tourist and day visitors' spending on fashion and other retail items contributed a further $4.5 billion, bringing total spending to $14.6 billion, or 31 per cent of total spending in these categories.
4. BIDs are designated commercial districts where property owners agree to pay a self-imposed tax to finance services such as street maintenance and beautification, public safety measures, and promotional and marketing activities.
5. Developments in the South Bronx also illustrate some of the drawbacks of urban ethnic tourism as many visitors seem less attracted by the community's complex social history and cultural diversity than by its image as a fearsome symbol of urban decay. At least some commercial tour operators exploit the stereotypes associated with the South Bronx for their own monetary gain.
6. Sandford (1987) in her study on tourism in Harlem questioned this characterisation, arguing that tour operators were trying to challenge existing stereotypes and paint a realistic picture of Harlem. Sandford concluded that Harlem's blight—not its culturally and historically important sights—was the main reason for most tourists of the time to venture into the neighbourhood.
7. The EZ's approach to neighbourhood revitalisation draws upon Michael Porter's argument that inner-city neighbourhoods should capitalise on their 'true' competitive advantages to alleviate poverty and reduce welfare and government aid dependency. In the case of Harlem, entertainment and tourism were, along with retail, health and business services, identified as 'high growth industries' (Gitell 2001; Hoffman 2003).
8. The removal of 125th Street's eye-catching street vendors in 1994 was a case in point. Their disappearance cost the street much of its tourist appeal but was according to Kelley (2007: 67) an important prerequisite to prepare the way for big capital: 'Although conflict between 'legitimate' businesses along 125th Street and the street vendors had been brewing since the 1970s, it is not an accident that the first military operations against them coincides with initiatives to woo Gap and Starbucks into opening shop in Harlem'.

REFERENCES

Audience Research and Analysis (2000) *Upper Manhattan Tourism Market Study: A Study of Visitors to Upper Manhattan Including their Economic Impact and Local Spending*. Prepared for the Upper Manhattan Empowerment Zone Development Corporation, Inc., December.

Brooklyn USA (2004) 'Marty hosts tourism initiative'. <http://www.brooklyn-usa.org/Pages/OIB/OIB_04/Jan2004.htm> (accessed 14 June 2005).

Chira, S. (1989) 'Once-wary Japanese now seek Harlem's soul', *New York Times*, 5 July: A1.

Chinyelu, M. (1999) *Harlem Ain't Nothing but a Third World Country*. New York: Mustard Seed Press.

Cocks, C. (2001) *Doing the Town: The Rise of Urban Tourism in the United States*. Berkeley: University of California Press.

Collins, G. (1998) 'Touting the Flavor of the Big Apple', *New York Times*, 9 September: B2.

Conforti, J. (1996) 'Ghettos as tourism attractions', *Annals of Tourism Research*, 23(4): 830–42.

Dávila, A. (2004) 'Empowered culture? New York's Empowerment Zone and the selling of El Barrio', *The Annals of the American Academy of Political and Social Science*, 594(1): 49–64.

Dowling, R. (2007) *Slumming in New York: From the Waterfront to Mythic Harlem*. Chicago: University of Illinois Press.

Eligon, J. (2007) 'An old record shop may fall victim to Harlem's success', *New York Times*, 21 August: B1.

Fainstein, S. (2007) 'Global transformations and the malling of the South Bronx', in Hammett, J. and Hammett, K. (eds) *The Suburbanization of New York*. New York: Princeton Architectural Press, 167–77.

——. (2005) 'The return of urban renewal', *Harvard Design Magazine*, Spring/Summer (22): 9–14.

Fainstein, S., Hoffman, L. and Judd, D. (2003) 'Making theoretical sense of tourism', in Fainstein, S., Hoffman, L. and Judd, D. (eds) *Cities and Visitors: Regulating People, Markets and City Space*. Oxford and Cambridge: Blackwell, 239–53.

——. (eds) (2003) *Cities and Visitors: Regulating People, Markets, and City Space*. Oxford and Cambridge: Blackwell.

Fainstein, S. and Powers, J. (2006) 'Tourism and New York's ethnic diversity: An underutilized resource?', in Rath, J. (ed.) *Tourism, Ethnic Diversity and the City*. London and New York: Routledge, 143–163.

Gates, Jennifer A. (1997) *Strangers in New York: Ethnic Tourism as a Commodity, Spectacle and Urban Leisure in Three Manhattan Neighborhoods*. PhD dissertation, New York University.

Gilbert, D. and Hancock, C. (2006) 'New York and the transatlantic imagination: French and English tourism and the spectacle of the modern metropolis, 1893–1939', *Journal of Urban History*, 33(1): 77–107.

Gitell, M. (2001) *Empowerment Zones: An Opportunity Missed. A Six-City Comparative Study*. Research report published by the Howard Samuels State Management and Policy Center at the Graduate School and the University Center of the City University of New York.

Greenberg, M. (2008) *Branding New York: How a City in Crisis was Sold to the World*. London and New York: Routledge.

Greene, L. (1979) *Harlem in the Great Depression, 1928–1936*. PhD dissertation, Columbia University.

Gross, J. (2005) 'Business Improvement Districts in New York's low and high income neighborhoods', *Economic Development Quarterly*, 19(2): 174–89.

Hammett, J. and Hammett. K. (eds) (2007) *The Suburbanization of New York*. New York: Princeton Architectural Press.

Hoffman, L. (2003) 'The marketing of diversity in the inner city: Tourism and regulation in Harlem', *International Journal of Urban and Regional Research*, 27(2): 286–99.

Hostetter, M. (2002) 'The cultural economy', *GothamGazette.com*, 1 January 2003. <http://www.gothamgazette.com/article/arts/20030101/1/35> (accessed 14 June 2005).

Hughes, L. (1940) *The Big Sea*. New York: Hill and Wang.

Johnson, J.W. (1991) *Black Manhattan: Account of the Development of Harlem*. New York: Da Capo.

Kelley, R. (2007) 'Disappearing acts: Harlem in transition', in Hammett, J. and Hammett, K. (eds) *The Suburbanization of New York*. New York: Princeton Architectural Press, 63–73.

Lin, J. (1998a) 'Globalization and the revalorizing of ethnic places in immigration gateway cities', *Urban Affairs Review*, 34(2): 313–39.

———. (1998b) *Reconstructing Chinatown: Ethnic Enclave, Global Change*. Minneapolis: University of Minnesota Press.

MacCannell, D. (1999) *The Tourist: A New Theory of the Leisure Class (Revised Edition)*. New York: Schicken.

Marcuse, P. (1998) 'Space over time: The changing position of the Black ghetto in the United States', *Netherlands Journal of Housing and the Built Environment*, 13(1): 7–22.

Maurrasse, D. (2006) *Listening to Harlem: Gentrification, Community and Business*. London and New York: Routledge.

Miles, S. and Miles, M. (2004) *Consuming Cities*. New York: Palgrave MacMillan.

New York Times (1923) 'Save the sacred times', 15 August: 16.

———. (1912) 'New York entertains nearly 200,000 strangers daily', 18 August: SM4.

———. (1884) 'Slumming in this town', 14 September: 4.

Newman, K. and Wyly, E. (2006) 'The right to stay put, revisited: Gentrification and resistance to displacement in New York', *Urban Studies*, 43(1): 23–57.

Oakley, D. and Tsao, H.S. (2006) 'A new way of revitalizing distressed urban communities? Assessing the impact of the federal Empowerment Zone program', *Journal of Urban Affairs*, 28(5): 443–71.

Office of the New York Comptroller (2006) *Economic Notes* 14(3). New York: Office of the New York Comptroller.

Oppenheim, C. (1978) 'Escorts provided: Harlem seeks a tourism future', *Chicago Tribune*, 29 October: 48.

Osofski, G. (1963) *Harlem, the Making of a Ghetto: Negro New York 1890–1930*. New York: Harper and Row.

Pogebrin, R. (2005) 'Groups vie to reimagine historic theater in Harlem', *New York Times*, 1 February: E7.

Rath, J. (2005) 'Feeding the festive city: Immigrant entrepreneurs and tourist industry', in Guild, E. and van Selm, J. (eds) *International Migration and Security: Opportunities and Challenges*. London and New York: Routledge, 238–53.

Rule, S. (1979) 'Harlem, tourist center in 20's and 30's, seeks rebirth', *New York Times*, 9 August: B3.

Sacks, M. (2006) *Before Harlem: The Black Experience in New York Before World War I*. Philadelphia: University of Philadelphia Press.

Sagalyn, L. (2001) *Times Square Roulette: Remaking the City Icon*. Cambridge: MIT Press.

Sandford, M. (1987) 'Tourism in Harlem: between negative sightseeing and gentrification', *The Journal of American Culture*, 10(2): 99–105.

Sante, L. (1991) *Low Life: Lures and Snares of Old New York*. New York: Vintage.

Sassen, S. (1991) *The Global City: New York, London, Tokyo*. Princeton and Oxford: Princeton University Press.

Savitch, H.V. and Kantor, P. (2002) *Cities in the International Marketplace*. Princeton and Oxford: Princeton University Press.

Severo, R. (1984): 'Harlem trying to become a "must see" on tour itineraries', *New York Times*, 15 August: B1.

Shaw, S., Bagwell, S. and Karmowska, J. (2004) 'Ethnoscapes as spectacle: Reimaging multicultural districts as new destinations for leisure and tourism consumption', *Urban Studies*, 41(10): 1983–2000.

Tracey, E. (1905) 'Passing of the Old Bowery: a retrospect', *New York Times*, 5 July: SM 4.

Williams, T. (2008) 'In Harlem, 2 record stores go the way of the vinyl', *New York Times*, 21 January: B1.

2 Ethnic Minority Restaurateurs and the Regeneration of 'Banglatown' in London's East End

Stephen Shaw and Sue Bagwell

INTRODUCTION

Brick Lane, a busy thoroughfare of Spitalfields in London's East End, has been a hub of religious, social and commercial activity for successive waves of immigrants for over three centuries. Over the past 30–40 years, the area has accommodated London's largest Bangladeshi population, one of the UK's poorest minorities. Initially, many Bangladeshis found work in the area's long-established textile industry. But Spitalfield's 'rag trade' was unable to keep up with global competition, pushing unemployment to new heights. From the mid-1970s, the street became the scene of periodic intimidation by right-wing race-hate groups; ugly images of these confrontations were communicated widely by the news media. Then, somewhat against the odds, a small cluster of cafés began to attract customers from the white majority culture as well as adventurous tourists. Bangladeshi landlords converted run-down commercial buildings into restaurants, and thus began a spectacular re-orientation of the local economy in the 1990s.

Over the past decade or so, state involvement in urban regeneration and wealth and job creation has helped stimulate a local bonanza in Asian-style cuisine catering to non-Asians. Less tangibly, it can be argued that prosperity and the promotion of a positive image for Brick Lane has lifted the self-esteem of a minority that only recently suffered severe poverty, a poor environment and racially motivated violence. By the late 1990s, this mix of economic and social goals seemed fully in tune with the incoming New Labour government's 'Third Way' agenda, a discourse that promised business-led regeneration combined with a new emphasis on social inclusion and capacity building: a shift from top-down government to participatory governance. Brick Lane thus became something of a showcase.

This chapter examines the circumstances that transformed Brick Lane into 'Banglatown—London's Curry Capital', focusing on the narratives of place promoted by new alliances of local government, business and civil society. While these partnerships widened participation, it would be naïve to assume that the 'stakeholders' enjoyed equal influence over decisions regarding public intervention, some of which had profound implications for the area and their own well-being. More specifically, this chapter examines a programme to package and sell a minority 'culture'—an exotic spectacle for the benefit of visitors more affluent than the local population—which has enabled Brick Lane restaurant owners to create new wealth and jobs. This approach is compared to alternatives that have developed in the revitalisation of two other thoroughfares associated with Asian communities—Green Street in Northeast London and Southall Broadway, near Heathrow in the West—both of which adopted a more pluralistic and hybrid vision of connections with contemporary Asian cultures.

NATIONAL-LOCAL STRUCTURES FOR INCORPORATION

The election of New Labour in 1997 followed nearly 20 years of Conservative government instilling the spirit of 'urban entrepreneurialism' into UK local authorities. Local authorities were not only expected to eschew bureaucratic practices and remove barriers to business-led regeneration; they were encouraged to become more entrepreneurial in their own right. As Ashworth and Voogt (1994) have observed, the principle that cities—and areas *within* cities—should compete against one another sat uncomfortably with the regional/urban planning that prevailed elsewhere in Europe. However, in the UK in the early 1980s, many urban areas faced rapid industrial decline and rising unemployment, with no prospects for state intervention to stem further losses. If they wished to attract/retain investors, developers, high-income residents, visitors and other desired groups, they now had little choice but to develop strategies of self-promotion, more typical of their North American counterparts (Ward 1998).

Local authorities were required to compete for central government grants under 'City Challenge', and from 1994, the 'Single Regeneration Budget' (SRB)—the regime that has nurtured the tourist economy in Brick Lane and other schemes discussed below. Local authorities had to make a convincing case that concurred with the government's emphasis on 'self-help'; councils had to enlist support from diverse agencies, such as developers, landowners, banks, hotels, trainers, cultural and community groups. At times unlikely bedfellows, differences between members of an SRB Partnership were generally played down. If successful, a small-area programme team would be set up, typically for five years. In areas of recent immigration and settlement, the acumen of minority entrepreneurs and their contribution to the city's 'cosmopolitan' tourism offering provided a compelling story-line of 'globalisation from below' that many interest groups and political factions could support in cities such as Bradford, Liverpool, Birmingham and London (Henry *et al.* 2002; Shaw *et al.* 2004).

Michael Porter's influential (1995) thesis that public intervention should work with the grain of market forces inspired developments in the UK, including the third round of SRB programmes, 'Building Business'

(1997–2002). For Porter, the relatively high cost of real estate, poor environment and infrastructure and crime and security problems in inner cities are disadvantages that must be addressed. However, the state should not respond to decline by throwing good money after bad. Rather, it should free up the economy and assist the true sources of competitive advantage, especially entrepreneurial talent within minority communities, a supply of low-skilled but motivated workers and proximity to downtown and entertainment areas. According to this argument, the enlightened self-interest of minority businesses can play a critical role in enhancing place competitiveness, in reconnecting local economies with city-wide and even global markets. All this can be presented as a potent statement of popular capitalism. At the same time, it can offer evidence of an inclusive, tolerant and cohesive society: a discourse more traditionally associated with the Left.

The incoming government was keen to promote its 'Third Way' agenda: a middle course between 'excessive statism' and laissez faire capitalism (Giddens 2000). Local authorities were expected to cultivate closer relationships with local business and 'third sector' organisations: non-governmental, 'value-driven' organisations that 'principally reinvest surpluses in the organisation or the community' (HM Treasury 2005: 7). The 86 most deprived areas that received Neighbourhood Renewal Funding were expected to develop Local Strategic Partnerships, a concept that was later extended nation-wide. The intention was that small area-based collaborations would foster a more 'joined-up' approach to tackle the multi-faceted problems of deprived areas. Far from being abolished, SRB continued through to 2007, retaining its key principles of competitive bidding and local collaboration. There was, however, a new emphasis on meeting *social* objectives—inclusion, capacity-building and wider participation—as well as those concerned with economic and physical regeneration.

More traditional structures of representative democracy were considered too bureaucratic to deal effectively with questions of identity in a multicultural and global/local world (Newman *et al.* 2004: 204), and often lacked the trust of local communities. As emphasised in the emerging body of governance theory, this shift from hierarchical government to participatory local governance has created new opportunities for actors previously excluded from decision-making (Taylor 2007: 297). In the case of the SRB programmes, Partnerships were required to demonstrate how they involved the 'community' in preparing the bid and in arrangements to facilitate wider participation. According to government guidance, this was 'likely to include the faith-based voluntary sector . . . ethnic minorities and local volunteers' (DETR 1998: 5). But as Edwards (2003) points out, in practice it was largely a matter for the Partnership to decide who or what the local community might be, who should be consulted, and what form this might take. On its re-election in 2006, New Labour re-affirmed its broad commitment to community engagement to improve public service delivery at the neighbourhood level. Nevertheless, critics point to the imbalance of power and resources that persist within local communities. The 'invited spaces' created through 'beyond-the-state' governance may thus privilege certain actors at the expense of others less able to play the system (c.f. Cornwall 2004; Swyngedouw 2005; Taylor 2007).

FROM RAG TRADE TO RICHES?

The regeneration of Brick Lane highlights tensions over the development of the tourist economy and the privileging of particular voices. The case study thus illustrates both the strengths and weaknesses of the system of economic and political incorporation described above. By the mid-1990s, the area's deprivation demanded urgent attention: further decline of the historic 'rag trade' of textile workshops and wholesaling was exacerbating already high levels of unemployment, compounded by poverty, poor housing conditions and the racist abuse experienced by many recent immigrants from Bangladesh. Some Bangladeshi entrepreneurs had nevertheless acquired commercial premises, and were responding to a market for cafés and curry restaurants that attracted white as well as Asian customers. And as a new influx of artists and designers came to live and work in Spitalfields, it gained a fashionably counter-cultural atmosphere that appealed to many young urban professionals.

The London Borough of Tower Hamlets, the area's local authority, began to consider the potential of leisure, tourism and hospitality as stimuli for economic and physical regeneration. Together with two adjacent inner-city boroughs, city institutions and voluntary organisations, it joined a partnership led by the City of London Corporation (Shaw and MacLeod 2000; Shaw 2007a). Known as the 'City Fringe Partnership', the consortium applied for SRB funding for proposals that included development and promotion of the historic neighbourhoods outside the Square Mile of the City of London as Emerging Cultural Quarters (City Corporation 1996: 17):

> These cultural areas, unique to the capital and on the doorstep of the City, will be developed to provide a resource for tourists as well as employees and business visitors, helping to enhance the City's reputation as the premier European business location.

Indeed, from medieval times, this swathe of settlements just outside the city walls had provided a home to marginalised groups and institutions whose presence was unwelcome within. About half an hour's walk north from the Port of London, Spitalfields became home to newly arrived immigrants, many fleeing religious persecution and/or poverty in their homelands. Protestant Huguenots expelled from France in the late seventeenth century established silk weaving, a luxury commodity for London's elite (Museum of London 1985; Shaw and Karmowska 2004). As their skills grew redundant with industrialisation over the next century, most moved away but others took their place.

The association of the area with textile manufacturing and wholesaling continued as Brick Lane became a hub of the Jewish East End following the exodus from central and northern Europe in the late nineteenth and early twentieth century. When this population moved away to higher income suburbs in the 1960s and 1970s, Bangladeshi entrepreneurs acquired the businesses, while many of their compatriots from Sylhet came to work in the 'sweatshops'. The historic contribution of immigration to the life of the city, as well as the exotic culture of its contemporary residents, was emphasised in the successful application for state funding made by the City Fringe Partnership (City Corporation 1996: 5–6):

> the cultural diversity and strong entrepreneurial culture have produced a strong base in leisure facilities, entertainment and the arts all within a short walk of the City. On offer is an array of restaurants, including the well-known tourist attraction Brick Lane, ethnic shops and thriving markets

While the street was now featuring in mainstream guidebooks, *The Rough Guide to London* (Humphreys 1997: 232) reveals its darker history:

> each step is accompanied by the smells of spices from the numerous cafés and restaurants, the bright colours of the fabrics which line the clothes shop window, and the heavy beat of Bhangra music from the shops and passing cars . . . hidden behind this façade, though, are crowded flats and sweatshops that would not look out of place in Victorian times, and a history of racism that stretches back centuries.

In more recent times, Brick Lane has gained a celebrity status that has placed it even more firmly on the tourist map, with the world-wide popularity of Monica Ali's *Brick Lane* (2003), released in film in 2007. Other literary celebrations of the street and its rich social histories include *On Brick Lane* by Rachel Lichtenstein (2007) and *Salaam Brick Lane: A Year in the New East End* by Tarquin Hall (2005).

Further away from the city centre, London's post-colonial suburbs have also become home to Asian communities. Two particular neighbourhoods have come to be known as 'quarters' for leisure consumption, though they have a much lower profile and have evolved very differently from Brick Lane. Green Street, six miles (nine km) east of central London, has come to be known as the 'Asian Bond Street' due to its growing number of Asian fashion and jewellery shops. Until the 1970s the local population was predominantly white working class; closure of the docks and several large local factories led to unemployment, and as the former group moved away, Asians and Afro-Caribbeans moved in. Key drivers in the revitalisation of Green Street were East African Asian entrepreneurs, expelled by the Ugandan dictator Idi Amin in 1972. Experienced traders, they began to open up Asian grocery and sari shops. The opportunity to capitalise on this trend was identified by the local authority, which sought to address the high levels of unemployment and deprivation that persisted into the 1990s. In 1994, a successful SRB bid led to a £8.5m regeneration programme. Its objectives were to make Green Street a centre of regional significance for the largely Asian but multicultural community, providing a new dimension to East London's economy (LB Newham 1994).

Southall, 11 miles (17 km) to the west, has the largest Asian population in London (63 per cent), the majority originating from the Indian Punjab. Close proximity to Heathrow Airport and work in local factories made the area attractive to immigrants in the late 60s and 70s. Unfortunately, Southall also became a focus for racial tensions, including disturbances and police action in 1979 that led to the death of teacher Blair Peach. Today, Southall has a thriving retail centre that claims to be the largest Asian shopping area in Europe, offering a colourful array of food and fabric shops, jewellers and restaurants. Nevertheless, its hinterland experiences high levels of deprivation and unemployment, enabling Southall to bid for government regeneration funds. But instead of exploiting popular notions of exotic leisure consumption to non-Asians, the strategy has been to promote Southall as London's 'Gateway to Asia'—not just for the area's Asian businesses, but for all those keen to exploit potential links with growing Asian economies.

CREATING STREETSCAPES OF CONSUMPTION IN BANGLATOWN

Returning to the main case study, a significant driving force that encouraged diverse groups to support the emerging visitor economy in Brick Lane was another SRB programme, carried out in parallel with City Fringe SRB. Known as 'Cityside' and led by LB Tower Hamlets, it was awarded £11.4 million in government funding (1997–2002) to 'strengthen links with the City and encourage diversification of the local economy', of which £1 million was allocated to 'Raising the Profile'. According to the bid proposal (LB Tower Hamlets 1996: 1), Cityside would 'pioneer a new model of regeneration' that would lift investor confidence. To encourage visitors, attention was focused on the main access points: new infrastructure funded through the two partnerships included 'Eastern' style gateways, signage and brighter street lamps incorporating 'Asian' motifs. At its southern entrance, the main approach from the City, it was important to ensure that affluent visitors would feel safe as well as welcome, especially after dark.

Shortly after its inception, Cityside set up a 'Town Management' group whose remit included two annual festivals, Baishakhi Mela in spring and the Brick Lane and Curry Festival in autumn. It was through this forum

LONDON

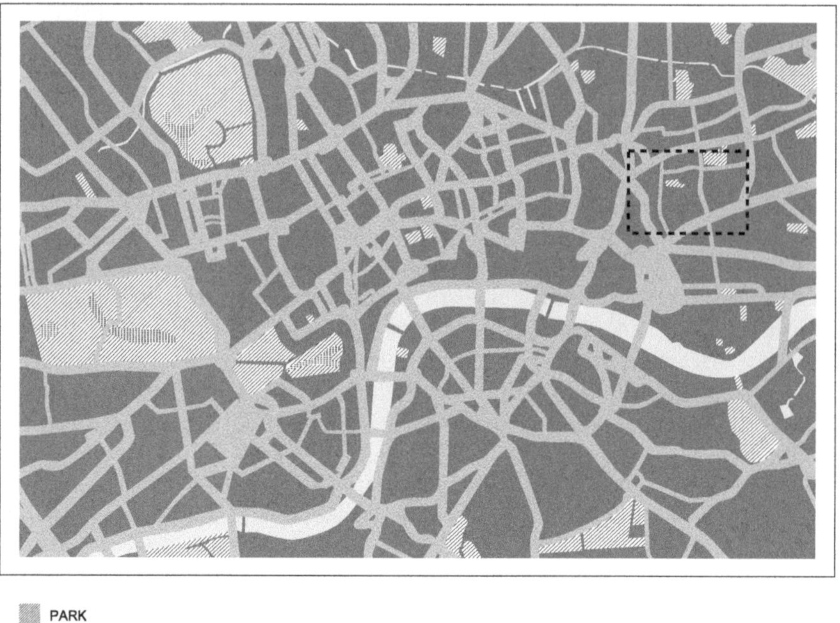

Figure 2.1 Location of Bircklane in London. *Map designed by Ana Sala.*

that 'Banglatown' emerged as the place-brand. Some traders wanted to create an even more striking environment with brightly coloured paving and street furniture. This re-presentation of the street for 'tourists', however, was not universally approved. By 2001, Baishakhi Mela was attracting 60,000 visitors. But as Eade (2006) observes, although the festival was promoted as a multicultural spectacle, some 'strict' Muslims frowned upon the music and dancing; there were particular sensitivities around places of worship on Fridays. Although supported by secular communities, the influential Imam of East London Mosque condemned the festival as an un-Islamic event that would lead young Bangladeshis astray.

Cityside's vision was nonetheless 'to achieve a quantum leap in the area's status as a visitor/cultural destination' (LB Tower Hamlets 1996: 13)—an aspiration that seemed ambitious to many, given the area's troubled past. In practice, the attraction of Asian food tuned to Western consumption greatly exceeded expectations as 'Banglatown' became one of London's best-known centres for ethnic cuisine. A survey carried out for Cityside noted that in 1989 there were only eight cafés/restaurants in Brick Lane, with a few additions in the early 1990s. Between 1997 and 2002, this rose to

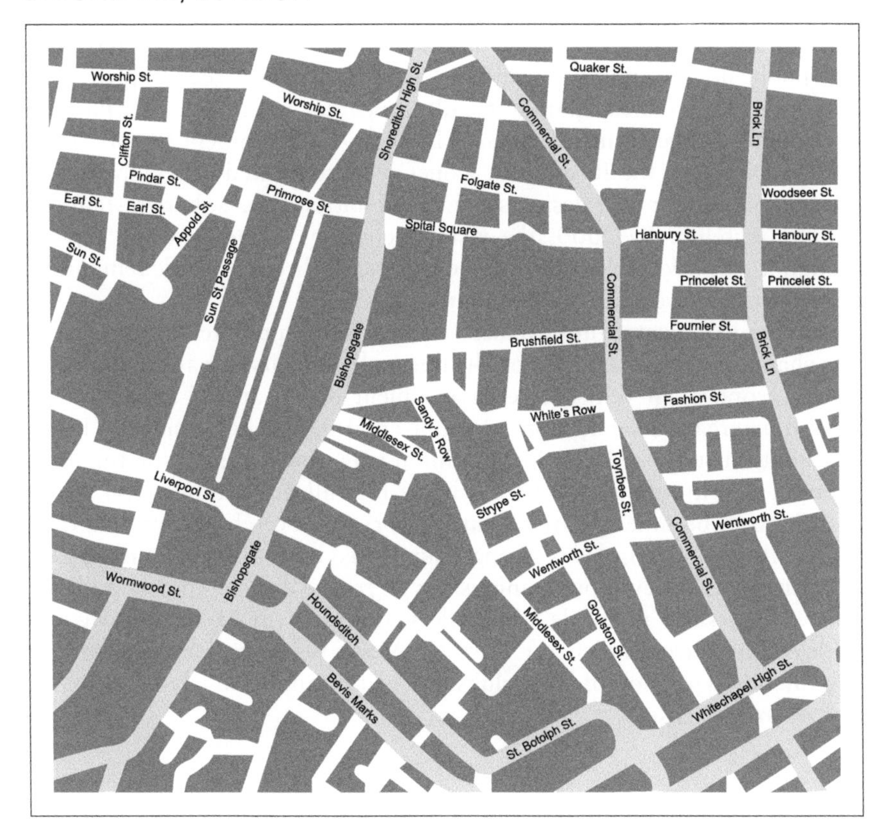

Figure 2.2 Map of Bricklane. *Map designed by Ana Sala.*

41, of which 16 had opened between 2000 and 2002, making Banglatown 'home to the largest cluster of Bangladeshi/"Indian" restaurants anywhere in the UK' (Carey 2002: 12). All reported that by the early 2000s their clientele was 'overwhelmingly white', with a clear majority (70 per cent) in the 25–34 age group, and predominantly male (*ibid.*: 4). The restaurants now employed around 400 staff, though most had difficulty recruiting suitable waiters due to the low pay and long, unsociable hours. At the time of writing (2008), there are nearly 60 restaurants; persistent on-street canvassing of customers by waiters offering discounts and free drinks suggests oversupply of an undifferentiated product.

This boom in Asian cuisine attuned to Western tastes has been stimulated, not only by public investment, but by the Borough Planning Committee's decision to permit the conversion of local shops into restaurants in the central section of Brick Lane (LB Tower Hamlets 1999). Overriding the approved development plan for the area, this strip was designated a 'Restaurant Zone' where applications for restaurants, cafés, hot food outlets, public houses and bars would be 'favourably considered'. Improvement grants were offered to upgrade façades, together with advice on business development. The former Truman's Brewery—a prominent historic landmark adjacent to the Restaurant Zone—was renovated and converted to accommodate over 250 design studios, exhibition spaces, bars/restaurants and a nightclub. By 2000, the visitor economy was expanding rapidly, though it was acknowledged that the volume of through-traffic remained a severe constraint on further development.

An 'Environmental Improvement Area' scheme, prepared in response to this challenge, was publicised by the borough and Cityside as an uncontroversial proposal. A vehicle-free environment would attract more visitors, as well as create a safer, more pleasant environment for residents. Initial consultation suggested that local opinion was in favour; consultants were appointed to draw up a detailed scheme for full pedestrianisation. However, it soon became apparent that support for the scheme was far from universal. The proposal was strongly supported by a group of restaurateurs who anticipated new opportunities for *al fresco* dining. Understandably, other firms, such as textile wholesalers, opposed the exclusion of goods vehicles for collection/delivery of stock. Many residents were also against any further expansion of bars, clubs and restaurants, and late-night extension of their licenses.

POLITICS OF CULTURE AND CONTESTED VISIONS FOR BANGLATOWN

Unfortunately, a stormy town hall meeting led to physical blows and the police being called in. In the aftermath, consultants were able to persuade their clients that permanent closure of the street would be undesirable and unworkable. Moreover, the diverse local communities that lived, worked, owned businesses, shopped, socialised, worshipped and enjoyed their leisure time in Brick Lane had to be actively involved in the scheme's design. The two-year programme of consultation and participation that followed (2000–02) obtained a much more balanced representation of local opinion. Those who took part in the questionnaires, interviews, workshops, focus groups and drop-in sessions included residents from the socially/ethnically diverse neighbourhoods around Brick Lane, restaurant and other business owners, community organisations, police, public servants and visitors (Shaw 2007b).

As anticipated, there was a wide diversity of opinion. Nevertheless, when funding eventually became available (from Transport for London), the scheme that was implemented (2002–06) was guided by broad agreement over solutions to minimise conflicts between pedestrians and traffic, bars/restaurants and other firms, visitors and local residents. As a result, the street is now safe to cross at most points; drop-kerbs make Brick Lane more accessible for the mobility impaired and for parents with pushchairs, while allowing access to commercial vehicles. It has also been accepted that Banglatown's image is to be reflected through public art and non-permanent features such as street decorations rather than through permanent, territorial expressions of ethnic identity. When the work of Cityside was completed in 2002, the partnership could report outcomes that were very positive in terms of its remit. The project had prioritised 'Developing visitor attractions, improving the environment and supporting events, all designed to increase the number of visitors to the area who will spend money within it' (Cityside SRB3 2002: 5). This complemented other objectives including helping local firms develop trading links with the City/elsewhere, access to jobs, and encouraging growth in small businesses. Cityside's Final Report concluded (*ibid.*: 7):

> 'London E1', 'Spitalfields' and 'Banglatown' are now the regular focus of media attention. . . . The Raising the Profile programme has led to the successful creation of annual events such as Alternative Fashion Week, Baishakhi Mela, Brick Lane Festival and many other community events which draw increasing number of visitors to the area.

Indeed, by 2004 the Brick Lane Festival was warmly endorsed, not only by the mayor of London, but by the prime minister, who praised its contribution to 'a truly inclusive Britain that takes pride in its diversity' (Brick Lane Festival 2004: 2). As the Official Guide (*ibid.*: 3) proclaimed:

> The Brick Lane Festival captures the flavour and excitement of an area that has welcomed immigrants for over 2000 years [sic]. From its community led roots in the summer of 1996 it now promotes the attractions of Brick Lane and Banglatown to around 60,000 people from all over London and the UK.

Nevertheless, there were anxieties that the wider Banglatown project was creating something of a tourist bubble (c.f. Judd 1999). The Borough Planning Department grew concerned that far from narrowing the gap, spatial inequalities had widened. From 2002, LB Tower Hamlets has protected a section of Brick Lane to the south as a 'Local Shopping Parade' to contain any further spread of the Restaurant Zone; in practice, only one 'local' grocery store remains (Shaw *et al.* 2004: 1992). Furthermore, there was already evidence of conflict over the use of public space. Planning Officer Andrea Ritchie (2002) reported from a focus group facilitated by the borough:

> Older Bengali women stressed the point that they had to be escorted by their husbands and that they could not walk along Brick Lane at all

> because there are just too many men there, with all the visitors and [restaurant] staff. So, although it is their area, they are socially excluded from it.

Today, visitors who tend to dominate the street, by day and especially by night, are mostly non-Asian. They mingle with the designer-artists working in the studios and the young professionals living above the restaurants in new apartments. It has grown increasingly difficult to answer the criticism that whatever Brick Lane has to offer, it has little relevance to the everyday lives of the Bangladeshi community, except those for whom it represents a business proposition. Furthermore, there are anxieties over drug dealing, prostitution and street crime, especially on the boundaries between Brick Lane and an adjacent social housing estate, a concern that was amplified in a recent set of interviews with 25 traders (Carey and Ahmed 2006). The survey also highlighted unease over the (temporary) downturn in trade following the London tube bombings on 7 July 2005; some believed that the 'Banglatown' brand might well be problematic in the event of any terrorist outrage before or during the 2012 Olympic Games.

ALTERNATIVE MODELS FOR EMERGING 'ASIAN QUARTERS' IN LONDON

Despite their concentrations of Asian shops and restaurants, both Southall and Green Street have resisted the temptation to link place-branding to any one ethnic group. In the case of Green Street, rather than creating a 'curiosity', SRB funding was used for streetscape improvements celebrating the area's diversity (LB Newham 1994: 2). These included a pavement mural designed with local schools and community groups, and lamp post decorations and street furniture to symbolise 'togetherness'. A statue of the famous footballer Bobby Moore was erected to acknowledge the area's white working class roots and associations with West Ham Football Club, while sponsored cultural events included the Asian-inspired Runga Rung festival: a winter celebration with live music and fireworks, and a range of 'multicultural' events including Afro-Caribbean music, tea dances for the older white British community and more recently, East/Central European music.

The promotion of Green Street as a visitor destination only began in 2001, towards the end of the SRB programme. *This is Green Street* featured branded merchandising, advertising in magazines and on ethnic minority radio, and familiarisation tours for the national press and travel operators. The *Quality Endorsement Scheme* identified retailers providing high quality merchandise and exceptional service. Shopkeepers have also invested in major improvements. The consultants evaluating the SRB programme described changes to the street as 'phenomenal', emphasising the contribution of younger, innovative Asian traders prepared to invest in window displays and store layouts to improve the shopping experience (EDAW 1997). Today, Green Street is widely regarded as a showcase for new Asian British designers offering clothes and jewellery that fuse traditional Asian and Western influences. There are few vacant premises and the Planning Authority is considering an extension to the area designated for commercial use.

For some, in particular the owners of the more exclusive Asian fashion houses (towards the northern end of the street), the image building has not gone far enough. According to Shirley Coote (2007), a freelance consultant who supports small retailers, many felt that the new street lighting lacked distinctiveness and the promotional flyers did not reflect the upscale nature of their shops. While the council has continued to upgrade the streetscape, in 2003 a proposal to redevelop the popular but very dilapidated Queen Street market (at the southern end of Green Street) met fierce opposition. Traders from all ethnic backgrounds joined forces with residents to oppose proposals endorsed by the mayor to sell the site to developers. As a result, the market was saved and continues to be an important source of employment as well as affordable multi-ethnic food and household goods. It appears that the racial tensions of former times have largely faded; there remains a balanced provision of everyday food and household goods to meet local residents' needs and luxury fashions that attract wealthier Asian customers from across London and South East England.

The growth in shoppers and visitors from outside the borough led to such concern over traffic and parking that restrictions were introduced, though some shop owners complained that their stringent enforcement adversely affected trade. Nevertheless, despite the place-promotion, the street's fame seems largely restricted to Asian communities. All this may change with the development of the Olympic site for the 2012 London games, as LB Newham will be the host authority for Olympic Park, Stratford. While the borough is keen to exploit the business opportunities that the Games would bring, the survey by Carey and Ahmed (2006) discussed above suggests that some Green Street traders are unconvinced of the benefits. As in Brick Lane, some express fears that large retail and leisure chains will establish outlets in close proximity. West Ham Football Club has also announced its intention to relocate, thus threatening the survival of the remaining working class cafés, pubs and eel and pie shops near the ground, symbols of traditional East End life.

Since the 1970s, Southall Broadway in LB Ealing has been an important shopping (convenience, clothing) and entertainment (especially restaurants, cinemas) hub for Asian communities in West London and further afield. It attracts a different crowd from Green Street: those seeking the 'Southall experience' who may buy specific items such as spices. Non-Asian visitors do not dominate the area. The Himalaya Palace cinema shows Bollywood blockbusters; Diwali celebrations (processions, fireworks and entertainment) attract thousands. 'Little India' walking tours are run by a well-known Indian cookery writer who takes visitors to local temples, Glassy Junction Pub (where customers can buy pints with rupees) and around the shops and markets, explaining the ethnic foodstuffs and the meaning of local festivals (Czerniawski 2005).

In the late 1990s the *British Panjabis* (a voluntary organisation promoting the minority's language and culture) launched a campaign to brand the commercial area as 'Panjabi Bazaar'. Over 2,000 shoppers, shop owners and local MPs reportedly signed a petition proposing bilingual street signs, Asian-inspired street furniture and a handicraft market to sell specialist souvenirs to tourists. Campaigners, however, argued over how 'Punjabi' should be spelt; there were also fears that links with the Sikh community would divide the area along religious lines. A recent survey of local residents and businesses revealed scant support for such place-marketing, particularly among the younger generation (Verma 2007). And as community leaders point out, Southall has a growing Somali and Afghani population—to develop a mono-cultural brand would not reflect the dynamic nature of the community (Verma 2007; Bains 2008).

SRB nevertheless funded street decorations with Asian motifs, including neon chillies, and streetscaping with stylised pavement murals. Compared to Brick Lane, these are discreet, with benches that seem designed and positioned primarily for local residents to rest and converse. More recently, Southall Regeneration Partnership Board, the body that oversees the area's regeneration plans, has been keen to promote a more inclusive and contemporary international image for the area. The extended partnership—renamed 'Heathrow City Partnership' in 2003—has established itself as a more broadly based agency that will outlive the SRB-funded schemes; with funding from the London Development Agency, it gathers diverse community and business stakeholders. These include some of the UK's most successful and well-known businesses: multi-million pound enterprises such as Noon Products Ltd. and Sunrise Radio, symbols of a very different type of Asian enterprise to those visited on the Little India tours. The Greater London Authority (2002) has cited the board as a model of multi-racial co-operation that has helped transform Southall's previous reputation as a flash point for racial tension.

The partnership's current strategy emphasises the area's important global links. The *Southall Town Centre Strategy 2002–12* (Southall Regeneration Partnership 2002: foreword) envisages that '[b]y 2010 Southall will be an international gateway for excellence in multiculturalism and commercial development'. It thus seeks to foster and promote Southall's links to Asia. For example, it anticipates that the global popularity of Bollywood can be exploited by a range of local businesses, and encourages bilateral trade and networking between British and Asian businesses around the world. To support this interpretation of the Southall brand, the strategy suggests that the culture of the area's many communities should be further reflected in streetscape art and on high-quality gateways located at the town's entrance points. It suggests that a specific marketing strategy for Southall should 'increase awareness of the unique Asian retail offer to the pan-London and national market' (*ibid.*: summary). The divergences between these case studies mirror Ching Lin Pang's argument (in this book) that ethnic precincts may emerge and evolve along multiple trajectories and with very different outcomes in the post-industrial economy.

CONCLUSION

In all three case study areas, a handful of enterprising Asian businesses initiated investment to upgrade run-down commercial premises. From small beginnings, they created niche markets for ethnic cuisine, speciality retailing and other services that attracted customers from wealthier areas of the city and beyond. The stimulation of a visitor economy required co-ordinated action and public funding to re-image places long associated with the poverty of immigrant communities, and to make streets and public spaces more welcoming to casual strollers. From the 1990s to 2007, urban authorities in the UK were encouraged to form partnerships with commercial and third sector organisations, and bid for central government funds to pump-prime the regeneration of disadvantaged areas, including commercial streets in London's multicultural neighbourhoods. Where applications were successful, some ethnic minority traders and landlords found themselves in favourable positions. Individuals could benefit from schemes to assist business development while interest groups—notably restaurateurs—could help shape the local planning and policy-making framework.

There is good reason to conclude that such collaborations have encouraged an unprecedented level of strategic thinking and action to restructure local economies (c.f. Cullingworth and Nadin 2006: 371). From the late 1990s, the New Labour government targeted local initiatives that could meet their social inclusion criteria and demonstrate that the 'community' had been involved in formulating the vision. Nevertheless, the 'mix and match' approach (Jayne 2006: 196–98), with its emphasis on local responsibility for delivery, produced wide-ranging interpretations of the national Third Way agenda. 'Beyond-the-state governance' created new spaces for participation, opportunities for actors whose voices were seldom heard in the past (Somerville 2005; Swyngedouw 2005; Taylor 2007). However, within these 'invited spaces', some stakeholders made themselves more audible than others. The shift from more established systems of local democracy to intervention and funding through non-elected partnerships at times created circumstances that allowed well-connected players to capture the process and influence policy outcomes—perhaps to the detriment of others less able to make their views known.

Over the study period, the publicly funded programmes to regenerate Brick Lane, Green Street and Southall Broadway incorporated their respective interpretations of the government's priorities for regeneration and renewal. However, this was grafted onto a structure designed by the previous Conservative government to encourage a new spirit of competitiveness—the turn from urban managerialism to urban entrepreneurialism. State intervention that enabled favourably positioned firms to exploit the 'genuine competitive advantage' of their inner-city locations was advocated by Michael Porter (1995: 55–56) over a decade ago, and his arguments seem to have had a lasting influence in the UK and North America.[1] From this neo-liberal perspective, the disadvantages of the inner-city location must be identified and addressed; they may include run-down infrastructure, inadequate for business transactions and poor security (*ibid.*: 62–64). If these disadvantages can be overcome, a symbiotic commercial relationship may develop between inner-city neighbourhoods like Spitalfields and the brighter lights of the Central Business District. An accessible, safe and attractive enclave with a 'cosmopolitan' ambience to complement attractions in the nearby city centre and entertainment district—a multicultural neighbourhood transformed from a welfare-supported 'revenue sink' into an asset for the city, paralleling New York City's aspirations for Harlem discussed by Novy in this book.

In the competitive market for inward investment, visitors and government funding, there is considerable pressure to spin optimistic place-narratives for up-and-coming 'cosmopolitan' enclaves. Positive visions are formulated to 'raise the profile' while Cassandra voices are generally silenced to present a united front. Indeed, it can be argued that pragmatic urban entrepreneurialism tends to underplay social realities, the very problems that necessitate regeneration in the first place (Williams 2003: 24). If bids are successful, tensions may well re-emerge, not least over how the money is spent. This is well-illustrated in the case of Brick Lane, where a very public argument in the early 2000s erupted over proposals to further expand the 'tourist bubble'. While public engagement informed a very different scheme to improve the urban realm without overt territorial ethnic branding, important questions remain as to whether an urban visitor economy established through such agencies and national/local political structures bring lasting economic and wider social benefits to areas such as Spitalfields.

From the perspective of social policy, there is continuing anxiety that unacceptable levels of poverty and deprivation persist behind the colourful façade that this model deliberately sets out to create. In practice, it is questionable whether the desired harmony of economic, physical and social regeneration has been achieved. Cityside's *Final Report* (2002: 13) emphasised:

> the complete regeneration of an area is dependent on a much wider set of initiatives, particularly those that relate to social and leisure provision, the environment and safety. . . . Although we accept that the artificial split between economic and social programmes is far from ideal, Cityside Regeneration will be supporting Tower Hamlets Council to attempt to implement a joined up solution locally.

The re-imaging of Brick Lane, Spitalfields, has been remarkable. The shabby-*chic* of its New East End streetscape now features in most mainstream tourist guides to London. The popularity of its festivals is cited by policy-makers as evidence of a vibrant and inclusive multicultural society, in marked contrast to the violence against recent immigrants from Bangladesh in the 1970s and 1980s. Nevertheless, the stereotypical post-colonial images of the Indian sub-continent that prevail in the restaurant trade—Café Raj, City Spice, Le Taj and so on—seem a far cry from the higher goal of '*inter-cultural* cosmopolitanism', recently defined by Wood and Landry (2008: 93) as 'a capacity to recognise and engage with cultures other than one's own', or Sandercock's (2006: 37–38) hopes for cities in which people will live 'alongside others who are different, learning from them, rating new worlds with them'. Bloomfield and Bianchini (2004: 12) contrast the multicultural model, where state support is directed within well-defined boundaries of recognised cultural communities, with the 'pluralist transformation of public space, institutions and civic culture', an approach 'to facilitate dialogue, exchange and reciprocal understanding between people of different cultural backgrounds'.

The Partnership for Southall is now designated a 'City Growth' area: a very different, longer-term strategic approach. Though again based on business-led regeneration, it has a wider remit (and a wider geographical scope) than previous SRB schemes. Furthermore, Southall and Green Street have adopted approaches to the development of minority commercial enclaves that look beyond what Jacobs (1996: 100) has characterised as 'racialised construct[s] tuned to multicultural consumption'. Both have embraced a variety of cultures and seek a wide set of benefits for businesses, residents and community organisations. An optimistic view is that these partnerships are helping to establish quarters that will stand the test of time, able to evolve as residents with other identities move in. They certainly promote a more contemporary image of local communities: Green Street by promoting cultural fusions in fashion and jewellery, Southall through its proximity to Heathrow and its global links to Asian superpowers—images more likely to appeal to second and third generation Asian communities as well as to the varied tastes of non-Asian visitors.

NOTES

1. For example, Professor Porter stressed the need for business engagement and private sector leadership at the launch of 'City Markets', Centre for Cities, Institute for Public Policy Research in London, June 2006.

REFERENCES

Ali, M. (2003) *Brick Lane*. London: Doubleday.

Ashworth, G. and Voogt, H. (1994) 'Marketing and place promotion', in Gold, J. and Ward, S. (eds) *Place Promotion: The Use of Publicity and Marketing to Sell Towns and Regions*. Chichester: Wiley, 39–52.

Bains, H. (2008) Chairman, Southall Community Alliance, personal interview with the authors, London, 25 April.

Bloomfield, J. and Bianchini, F. (2004) *Planning for the Intercultural City, Intercultural City Series Book 2*. Stroud: Comedia.

Brick Lane Festival (2004) *Official Guide*. London: Ethnic Minority Enterprise Project.

Carey, S. (2002) *Brick Lane, Banglatown: A Study of the Catering Sector, Final Report*, Research Works Limited, Hendon, London, prepared for Ethnic Minority Enterprise Project and Cityside Regeneration.

Carey, S. and Ahmed, N. (2006) *Bridging the Gap: The London Olympics 2012 and South Asian-owned Businesses in Brick Lane and Green Street*. London: The Young Foundation and Agroni Research.

City Corporation (1996) *Revitalising the City Fringe: Inner City Action with a World City Focus*. London: City Corporation.

Cityside (2002) *SRB3 Final Report*. London: Cityside Regeneration.

Coote, S. (2007) Freelance consultant employed by Newham College to support small retailers, personal interview with author, Newham, London, 11 December.

Cornwall, A. (2004) 'New democratic spaces? The politics and dynamics of institutionalised participation', *IDS Bulletin*, 35(2): 1–10.

Cullingworth, B. and Nadin, V. (2006) *Town and Country Planning in the UK, 14th Edition*. London and New York: Routledge.

Czerniawski, O. (2005) 'A walking tour of West London's Little India', *Untold London*. <www.untoldlondon.org.uk/news/ART31479.html> (accessed 25 September 2007).

Department of the Environment, Transport and the Regions (DETR) (1998) *Single Regeneration Budget Bidding Guidance: A Guide for Partnerships (Round 5)*. London: The Stationery Office.

Eade, J. (2006) 'Class and ethnicity in a globalising city: Bangladeshis and contested urban space in London's "East End"', in Arvaston, G. and Butler, T. (eds) *Multicultures and Cities*. Copenhagen: Museum Tusculanum Press, 57–69.

EDAW (1997) *Green Street Partnership Interim Study, Final Report*. London: EDAW.

Edwards, C. (2003) 'Disability and the discourses of the Single Regeneration Budget', in Imrie, R. and Raco M. (eds) *Urban Renaissance? New Labour, Community and Urban Policy*. Bristol: The Policy Press, 163–80.

Giddens, A. (2000) *The Third Way and its Critics*. Cambridge: Polity Press.

Greater London Authority (2002) *Rebuilding London's Future*, Report of the London Assembly's Economic Development Committee, March.

Hall, T. (2005) *Salaam Brick Lane: A Year in the New East End*. London: John Murray.

Henry, N., McEwan, C. and Pollard, J.S. (2002) 'Globalisation from below: Birmingham—postcolonial workshop of the world?', *Area*, 34(2): 117–27.

Her Majesty's Treasury, Department of Trade and Industry and Home Office (2005) *Exploring the Role of the Third Sector in Public Service Reform*. <http://www.hm-treasury.gov.uk/media/3/E/vcs_thirdsector210205.pdf> (accessed 23 June 2008).

Humphreys, R. (1997) *London: The Rough Guide*. London: Rough Guides.

Jacobs, J.M. (1996) *Edge of Empire: Postcolonialism and the City*. London and New York: Routledge.

Jayne, M. (2006) *Cities and Consumption*. London and New York: Routledge.

Judd, D. (1999) 'Constructing the tourist bubble', in Judd, D. and Fainstein, S. (eds) *The Tourist City*. New Haven and London: Yale University Press, 35–53.

LB Newham (1994) *Green Street: A Regional Role in Tomorrow's East London*. London: LB Newham.

LB Tower Hamlets (1999) *Brick Lane Retail and Restaurant Policy Review*. London: LB Tower Hamlets.
———. (1996) *Eastside Challenge Fund Submission* (NB the name was changed to 'Cityside'). London: LB Tower Hamlets.
Lichtenstein, R. (2007) *On Brick Lane*. London: Hamish Hamilton.
Museum of London (1985) *The Quiet Conquest: The Huguenots 1685–1985*. London: Museum of London Publications.
Newman, J., Barnes, M., Sullivan, H. and Knops, A. (2004) 'Public participation and collaborative governance', *Journal of Social Policy*, 33(2): 203–23.
Porter, M. (1995) 'The competitive advantage of the inner city', *Harvard Business Review*, May–June: 55–71.
Ritchie, A. (2002) Planning and Community Liaison Officer, LB Tower Hamlets, personal interview with authors, Bow Road, London, 27 September.
Sandercock, L. (2006) 'Cosmopolitan urbanism: a love song to our mongrel cities', in Binnie, J. *et al.* (eds) *Cosmopolitan Urbanism*. London: Routledge, 37–53.
Shaw, S. (2007a) 'Cosmopolitanism and ethnic cultural quarters', in Richards, G. and Wilson, J. (eds) *Tourism, Creativity and Development*. London and New York: Routledge, 189–200.
———. (2007b) 'Hosting a sustainable visitor economy: messages from London's Banglatown', *Journal of Urban Regeneration and Renewal*, 1(3): 275–85.
Shaw, S., Bagwell, S. and Karmowska, J. (2004) 'Ethnoscapes as spectacle: reimaging multicultural districts as new destinations for leisure and tourism consumption', *Urban Studies*, 41(10): 1983–2000.
Shaw, S. and Karmowska, J. (2004) 'Multicultural heritage of European cities and its re-presentation through regeneration programmes', *Journal of European Ethnology*, 34(2): 41–56.
Shaw, S. and MacLeod, N. (2000) 'Creativity and conflict: cultural tourism in London's city fringe', *Tourism, Culture and Communication*, 2(3): 165–75.
Somerville, P. (2005) 'Community governance and democracy', *Policy and Politics*, 33(1): 117–44.
Southall Regeneration Partnership (2002) *Southall Town Centre Strategy (2002–2012)*. London: Southall Regeneration Partnership and London Borough of Ealing.
Swyngedouw, E. (2005) 'Governance, innovation and the citizen: the Janus-face of governance-beyond-the-state', *Urban Studies*, 42(11): 1991–2006.
Taylor, M. (2007) 'Community participation in the real world: opportunities and pitfalls in new governance spaces', *Urban Studies*, 44(2): 297–317.
Verma, A. (2007), Chief Executive of Heathrow City Partnership, personal interview with authors, Southall, LB Ealing, London, 1 October.
Ward, S. (1998) *Selling Places: The Marketing and Selling of Towns and Cities 1850–2000*. London: Spon.
Williams, G. (2003) *The Enterprising City: Manchester's Development Challenge*. London: Spon.
Wood, P. and Landry, C. (2008) *The Intercultural City: Planning for Diversity Advantage*. London: Earthscan.

3 Gateways to the Urban Economy
Chinatowns in Antwerp and Brussels

Ching Lin Pang

This chapter examines two Asian neighbourhoods or 'Chinatowns' in Belgium—in Antwerp (Pang and Hauquier 2006; Pang 2007) and Brussels (Hsu 2007)—through the theoretical lens of immigrant entrepreneurs in the urban political and symbolic economies. My main aim is to analyse 'Chinatown on display' as a place consumed by both Chinese and non-Chinese.

THEORETICAL MODELS

Waldinger *et al.* (1990) argued from an interactionist perspective that ethnic entrepreneurship emerges when opportunity structures interact with group characteristics. While this model inspired numerous scholars and stimulated compelling research in the 1980s and 90s (Portes and Manning 1986; Portes and Jensen 1987; Sanders and Nee 1987; Waldinger 1993; Logan *et al.* 2003), it perhaps too one-sidedly celebrated the agency of ethnic entrepreneurs and workers. Cast on the liberal economic model, the interactionist framework also largely overlooked the role of government. In response, Kloosterman and Rath (2001, 2003; Kloosterman *et al.* 1999) advanced the 'mixed embeddedness' approach, which considers both opportunity structures and the intervention of public institutions. Following their lead, this chapter examines the role of institutional actors—particularly at the local level—in supporting immigrant entrepreneurship and the 'thematisation' of ethnic precincts. In my analysis, local actors include both service providers within civil society and policy-makers.

Theoretical inspiration was also sought in the literature on urban political and symbolic economies. A vast array of works underlines the blurred nexus between culture and market due to an ever-changing and expanding consumer culture (Featherstone 1995; Hannerz 1996). The nature of commodities have transformed at a dizzying pace: 'consumables' are no longer only manufactured products. For seemingly insatiable postmodern urbanites, consumption now extends far beyond the range of daily products. While coffee, chocolates, breakfast cereals, contraceptives, shoes, etc.

have become more sophisticated, consumption has also spilled over into the realm of services and experiences.

Places can also be consumed. Zukin (1991, 1995, 1998; Zukin *et al.* 1998) draws our attention to the commodification of specific neighbourhoods, reshaped and restyled to whet the appetite of the postmodern urbanite. Others (Hannigan 1998) insist that the 'fun factor' was already evident in major American cities at the beginning of the last century. The 'fantasy city' is an attempt by leisure merchants to mix entertainment and commerce—witness 'shopertainment', 'eatertainment' and 'edutainment'. Yet in recent times the scope and nature of this fun factor has changed beyond recognition.

Much thought has been given to the transformation of ethnic neighbourhoods into ethnoscapes (Appadurai 1990) or ethnic precincts (Collins 2007) of leisure and consumption (Conforti 1996; Shaw *et al.* 2004; Rath 2007; Taylor 2000). 'Chinatowns' are exemplary cases of thematised ethnic precincts (see for instance Anderson 1990; Chow 1996; Christiansen 2003; Pang and Rath 2007; Yamashita 2003; Zhou 1992). Some go further in suggesting that 'Chinatown'—as 'a projection of Imperial China' initiated by white American architects (Shen 2007: 122)—is in fact a precursor of the current 'simcity' or 'fantasy city'. The first Chinatown in San Francisco blended in with the architectural make-up of its surrounding areas. It was only after the 1906 earthquake destroyed much of the area that the 'architecturally non-descript' Chinatown became a 'Chinese-themed' precinct based on the Imperial architecture found in northern China, in line with Western perceptions of 'the East' at the time.

Both ethnic precincts—Chinatown Antwerp and Chinatown Brussels—function as Chinese ethnoscapes for leisure and consumption and share some core markers. Yet in-depth readings reveal significant internal differences which merit closer analysis. These divergences can be clustered under three main headings: entrepreneurs' social and migration background; their embeddedness in local communities; and the status of their neighbourhoods within broader urban settings.

MAPPING THE SETTINGS

Chinatown, Antwerp

Antwerp (2003 population: 470,421) is the largest city in Flanders, the northern part of Belgium. In the first half of the fourteenth century, Antwerp was the leading trading and financial centre of Western Europe; its wealth was based on its seaport and the wool market. In the sixteenth century Antwerp once again experienced a golden age. Its glorious past notwithstanding, contemporary Antwerp likes to label itself 'a pocket-size metropolis'.

Although Chinese migration began in the second and third decades of the last century, substantive flows leading to settlement and the formation of communities took place only in the late 1950s and 60s. In contrast to traditional immigration countries, Chinese settlement in Western Europe was highly dispersed. The first restaurant owners catered to the majority group rather than to fellow Chinese (Pang 2003b); scattered settlement was hardly conducive to the formation of ethnic enclave-type Chinatowns. The clustering of Chinese small businesses, mostly restaurants and food stores, only took off in the mid-1970s, and while some of

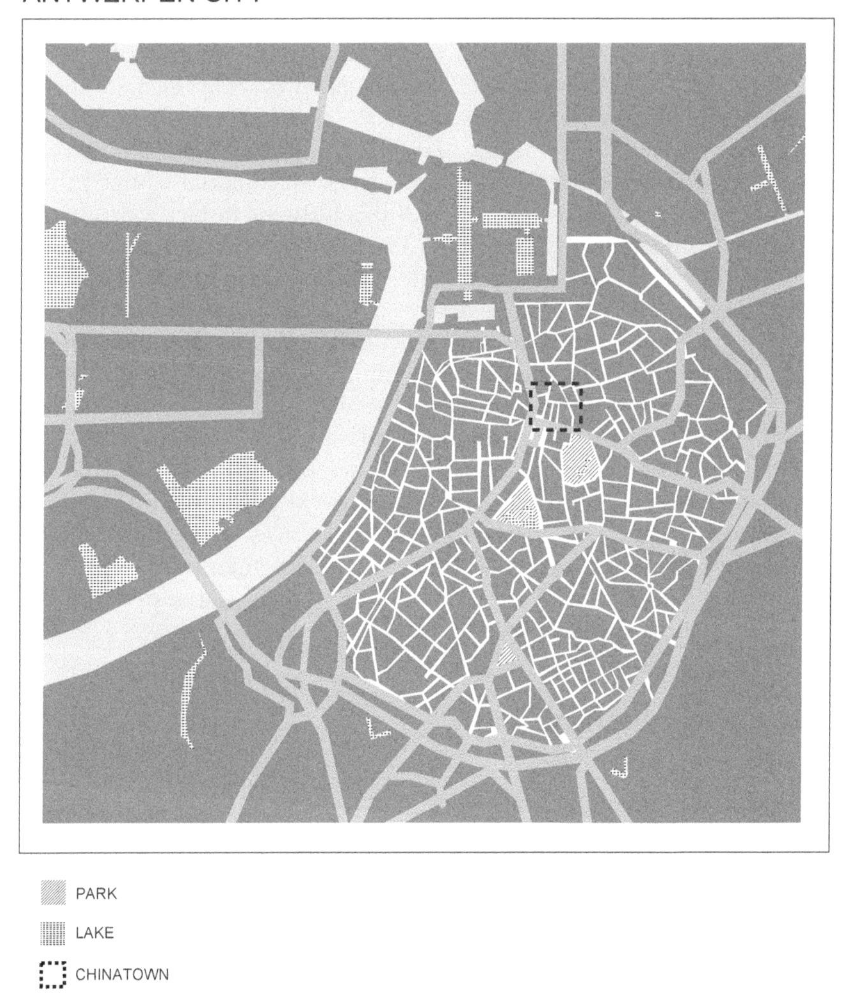

Figure 3.1 Location of Chinatown in Antwerp. *Map designed by Ana Sala.*

these first vestiges have vanished, others have remained. Three decades later, Chinatown Antwerp is a familiar ethnoscape along two main thoroughfares near Antwerp's Central Station: Van Wesenbeke Street and Van Artevelde Street.

The first stores in the area now commonly known as Chinatown opened in the mid-1970s. Despite their scattered settlement, the area around Antwerp Central Station became a meeting place for a growing group of Chinese immigrants who did their business shopping in 'the Criée' indoor market. Although their visits to the area were initially short and functional, many enjoyed the brief encounters with fellow Chinese. The fleeting exchange of greetings and information exchange sessions soon caught the attention of shrewd Chinese businesspeople who sought to provide these visitors with a meeting place in their restaurants. The embryonic formation of a Chinese social network quickened when a Surinamese-Chinese family from the Netherlands opened the Sun Wah supermarket in 1976.

Sun Wah attracted other Chinese businesses to its vicinity. The Chinese clientele grew significantly over the following decade as newly successful restaurateurs had disposable income. To display their newly acquired wealth and share their good fortune while embracing Chinese food culture, they visited Chinatown on their days off. Families would dine copiously every week—year in, year out—in restaurants serving southern Chinese cuisine in Chinese-style interiors, with Chinese spoken by customers, waiters and restaurant owners in an animatedly loud (*renao*) environment. While adults discussed their businesses, money matters and food, teenagers and children entertained themselves with magazines featuring the latest Hong Kong fads and gossip about local Chinese celebrities. Families expanded over time with the addition of in-laws, mostly Chinese but occasionally also a white person.

The first phase of the neighbourhood's thematisation was driven by the needs and preferences of Chinese immigrants, who came to regard the area as a diminished form of their home town. Embracing and celebrating their Chinese identity and descent through language, food culture, consumption of Chinese media and observance of the intricate system of Chinese inter-personal relationships, the neighbourhood became a space exuding 'Chineseness'. Many other businesses now opened their doors, including video rental shops, bakeries, travel agencies, dentists, doctors, acupuncturists and religious institutions. Social services and associations soon followed.

As the landmark of Antwerp's Chinatown, Sun Wah has become the largest Chinese supermarket in Belgium, specialised in a wide range of Asian foodstuffs. Products targeting the Subsaharan African population are a more recent addition, alongside the selection of *manga*, DVDs, Chinese and Japanese houseware and Chinese-style clothing. Sun Wah is now part of the Rolic Group, which distributes a free monthly newspaper discussing technology, financial news, culture and fashion in the Benelux countries.

Chinatown's two streets currently house 22 shops. They include 14 Chinese (mostly food) stores, two Thai supermarkets, one Chinese and one Thai barber, a Belgian bakery, a Belgian hotel, a Belgian snooker store and the indoor market, the Criée. The 15 restaurants include 12 Chinese, two Thai and one Japanese establishment. Incidentally, the Japanese restaurant was opened by a Chinese, while the Thai restaurants are run by Thais. In comparison to previous decades, the customers in Chinatown Antwerp today are a highly heterogeneous lot.

Chinatown/Asiatown, Brussels

As in Antwerp, most of Brussels' Chinese restaurants are scattered throughout the city. Nonetheless, there is a small cluster of restaurants at the edge of l'Ilot Sacré, the historic walking zone between la Grande Place (Market Plaza) and la Bourse (the stock exchange) in downtown Brussels. L'Ilot Sacré is not only renowned for its heritage but for its range of restaurants. Some guidebooks call the area 'the stomach of Brussels'.

Next to the l'Ilot Sacré is the Dansaert district, which includes Dansaert Street and Saint Goriks Square, with a prominent Chinese-Vietnamese/Asian presence. A handful of Chinese businesses on Dansaert Street target an exclusively Western clientele, while the streetscape of Saint Catherine Street is visibly marked by Chinese characters. The juxtaposition of Chinese characters with French and Dutch signs has not been contested by neighbourhood residents or the local government. The Chinese businesses include supermarkets, restaurants, gift and toy stores, a travel agency, a bakery, a barber and a traditional Chinese healing clinic. In comparison to Chinatown Antwerp, the development here is more recent, less homogeneous and indisputably more diverse.

Brussels' Chinatown is not the gradual outcome of Chinese community formation. The Kam Yuen supermarket attracts people from many ethnic and socio-economic backgrounds: Chinese, Vietnamese, Japanese, Koreans, the young and trendy living in the neighbourhood, EU nationals like the British, Dutch and French living in the posh municipalities outside the city, Belgian day trippers from outside the capital and others. The same mix of people can be found in most stores, their purchases often reflecting the tastes of their socio-economic and ethnic/cultural groups. A shopping trolley with fresh Chinese produce and large bags of rice most likely belongs to a Chinese or Asian person doing the family shopping. A shopping carrier with *krupuk* (shrimp crisps), sweet chili sauce, deep-frozen *dim sum* snacks and instant *ramen* noodles would be that of a young professional urbanite or a second generation Asian 'gone native'. Even in restaurants offering highly specialised foods—generally not enticing to the European palate—one can find non-Chinese clients. Hardly any store exclusively targets Chinese customers; some businesses, especially those on Dansaert's trendy thoroughfare, are frequented mainly by non-Chinese. Besides the

Chinese stores, there is a Thai supermarket, a handful of traditional Belgian restaurants, two popular fish shops, a Brussels hair salon, a pub and a night shop on the corner.

In contrast to the posh and avant-garde vibe of Dansaert Steet, Saint Catherine Street looks chaotic, *bric-a-brac* and shabby. In the same area, Saint Goriks Square—famous for its trendy bars, cafés and ethnic restaurants—hosts ethnic Chinese, Indian, Thai and Vietnamese restaurants and shops. The streetscape has a predominantly Asian feel. The Vietnamese and Thai restaurants were opened by Vietnamese of Chinese descent who arrived as refugees in the 1970s and 80s, among the first Asians who started small businesses in the neighbourhood.

BRUSSELS MAP

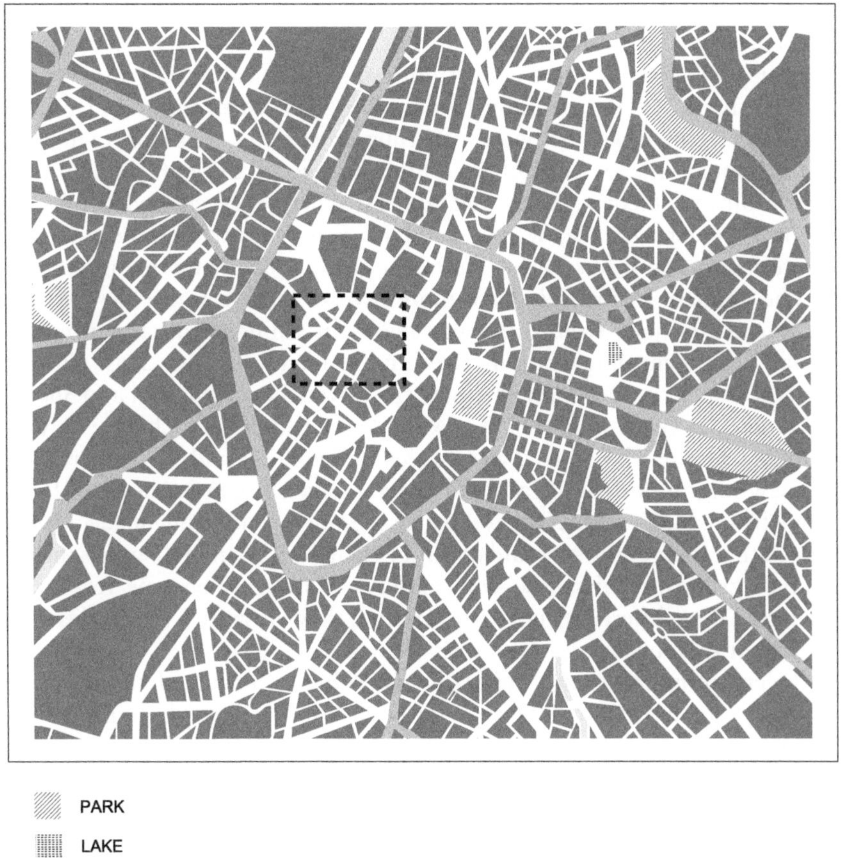

Figure 3.2 Location of Chinatown in Brussels. *Map designed by Ana Sala.*

DIFFERENT BACKGROUNDS BUT EQUALLY SUCCESSFUL

The involvement of Chinese immigrants in Europe in the catering business is well-documented (Pang 2003a; Pieke and Benton 1998; Rijkschroeff 1998; Song 1999; Watson 1977). In contrast to immigration countries with a critical mass of co-ethnics, first generation Chinese immigrant entrepreneurs in Europe catered to the majority group. From the 1980s onwards, Chinese entrepreneurs could also serve co-ethnics in the inner-city areas of Antwerp and Brussels, while the following decades witnessed a further diversification of their clienteles.

A closer examination of the entrepreneurs' migration and social backgrounds reveals significant differences. The Chinese in Antwerp hail mostly from Hong Kong. Hong Kong was hit by recession in the 1950s and 60s, with crisis engulfing rice cultivation in the New Territories. The political turmoil in China that culminated in the Great Cultural Revolution was also felt across the border. Migrants left their same-surname ancestral villages in search of economic opportunities and political stability; those who arrived in the 1960s generally belonged to the same family and lineage networks. Chinatown Antwerp can thus be seen as a sort of 'extension' of traditional villages in southern China. Ethnic networks facilitated their migration and incorporation in Flanders, where they entered the highly profitable catering trade. While Mandarin is now growing in prominence, Cantonese is still spoken by a significant group in Chinatown Antwerp.

The Chinese in Brussels have more diverse backgrounds: they arrived in different periods from different regions in China, and vary in educational attainment, migration motives and residential status. Most Chinese in French-speaking Belgium originate from Wenzhou, Qingtian and Shanghai. Some are twice-migrants, for example the Indonesian Chinese or Vietnamese Chinese who arrived as refugees in the 1970s and 80s. Motives for the new migration from Mainland China, which began in the late 1970s, range from study to marriage to business. Some former students decided to stay so that their children could continue their education in Belgian schools; a small group of Taiwanese who arrived as students became restaurant owners. These differences notwithstanding, they all share a strong desire to start their own business.

Recent immigrants are generally more highly educated; unlike their counterparts in Antwerp, they master French and/or English. In her ethnographic study of Brussels' Chinatown, Hsu (2007) portrays people who are not afraid of making new investments or learning new business skills. One couple owned a seafood wholesale business for seven years before starting a real estate company. Another entrepreneur was a Chinese teacher in a language school for over a decade; as he wanted to move into a cultural industry, he worked for two years in a German design company to learn about marketing. Together with a Belgian partner, he then opened a store named Rouge that sells trendy Chinese gadgets to *bobos* (bohemian bourgeois). Armed with human capital and years of experience, Chinese entrepreneurs in Brussels have developed sharp business acumen, which probably explains their success.

Despite their different migration and educational backgrounds, Chinese entrepreneurs in both cities were generally successful. When asked, they all refer to hard work, flexibility, good human (family or other) relationships and business acumen as the main ingredients of their success. Re-enacting and personifying the Chinese migration dream that began in the 19th century, their migration and incorporation narratives mirror the ingrained Chinese view of migration as an economic opportunity—a view that also underpins the Anglo-Saxon literature on the agency of the immigrant entrepreneur who made it 'from rags to riches'.

EMBEDDEDNESS IN LOCAL COMMUNITY LIFE

Chinatown Antwerp and the Importance of Associations

The role of Chinese overseas associations is well-documented and widely discussed in the literature. Chinatown Antwerp is no exception as it provides fertile ground for community life. There are six Chinese associations, one Protestant church, one Buddhist temple, two Chinese martial arts schools, one Chinese language school and one Nepalese association. The Chinese associations include the Association of Chinese Settlement in Belgium, Chinatown vzw, the Chinese Association of Merchants, the Association of the Chinese Elderly, the Chinese Women's Association, the Fujian Chinese Association and two acupuncture centres. Most organisations are run by entrepreneurs who share a common language and region of origin, and who have similar migration backgrounds.

Whether they desired it or not, Chinese associations became intermediaries between the Chinese community and local policy-makers; they now play an important role in connecting immigrants and their organisations to the structures and activities of urban renewal in the post-industrial city. Social policies to support impoverished inhabitants and to combat poverty and crime in decayed urban areas tap immigrants' cultural and financial resources as well as entrepreneurial inclinations. Alongside official public funding, the Antwerp city government raised funds from Chinese associations and individuals to re-arrange the streetscape.

Since the 1990s an increasing number of non-Chinese have found their way to Antwerp's Chinatown. Chinese businesses adapted by expanding their range of Chinese-Asian foods and exotic Asian products. They also discovered the growing interest among non-Chinese for traditional festivities such as the Chinese New Year.

In the 1970s, celebrating the Chinese New Year was an exclusively Chinese affair; nowadays it attracts a large and diverse public. The festival consists of a lion dance accompanied by ear-deafening firecrackers and stage performances including Chinese opera and dance. On the major thoroughfare and on the square, stands sell Chinese food and gadgets, and offer palm reading and calligraphy. In the evening, a formal party organised by the Association of Chinese Settlement in one of Antwerp's major performance halls brings together Chinese, locals, city officials and representatives of the Chinese embassy. Besides Chinese New Year, the Mid-Autumn Festival has also been celebrated over the past few years. Evolving from a traditional harvest festival, it has become a family celebration where all gather at a huge banquet, in the same spirit as Thanksgiving in the United States. The Mid-Autumn Festival has become a social event in Antwerp's Chinatown, as it has in many other cities outside China. Last but not least, religious celebrations such as the birthday of the Buddha are now open to the general public.

These public celebrations are free of charge and open to all, creating a temporary sense of social cohesion in an otherwise divided neighbourhood. Walks through Chinatown are offered by different associations, both local and Chinese, and may include visits to the Chinese school, the Buddhist temple and a martial arts school before ending with a Chinese meal. The local government promotes interaction between the neighbourhood's ethnic minorities by supporting these events.

Chinatown Brussels: The Absence of the Third Sector

The clustering of Chinese and Asian businesses on Saint Catherine Street and Saint Goriks Square notwithstanding, Chinese merchants in Brussels show no inclination of forming their own associations or joining existing ones. The two largest Chinese groups in Brussels hail from Qingtian and Wenzhou; following the general pattern of overseas Chinese, they have formed their own associations. Besides the lack of a common region of origin, the Brussels merchants' lack of interest in joining these associations derives from their educational backgrounds and language proficiency: they are able to fend for themselves without the assistance of intermediaries, whether they be compatriots or others. Participating in association activities is furthermore deemed too costly, time-consuming and inconvenient, with most meetings and parties taking place after midnight. As Chinese merchants are not involved in Dansaert's local management committee either, their voices often go unheard.

The installation of lions, arch gates and other visible markers of 'Chineseness' is not a major concern for Brussels' Chinese merchants. They manage their businesses on a highly individual basis, relying on personal financial, human and social capital (some are related as family members or are business partners). Most entrepreneurs are involved in more than one line of business. For example, Little Asia, an upmarket Vietnamese restaurant, was opened by the daughter-in-law of the owner of the Sun

Wah supermarket in Antwerp. She gained national stature by writing a Vietnamese cookbook, while her sister opened an Asian toyshop next to her restaurant. The owner of Kam Yuen supermarket is also the owner of the Chinese bakery across the street, while the owner of the barbershop has a restaurant on an adjacent street.

URBAN RENEWAL AND THE NEW SYMBOLIC ECONOMY

Antwerp's Chinatown: Benefiting from Neighbourhood Renewal

Antwerp's Chinatown occupies an area targeted for urban renewal since the closure of the Rex cinema complex in 1993 (City of Antwerp 2004). It is an area plagued by problems shared by many inner-city neighbourhoods: decaying buildings, unemployment, rampant crime, drug dealing, prostitution, safe houses for the victims of human trafficking, etc. In 1994 the city drafted an outline for the renewal of the Central Station area, to be funded by the city and the Flemish (Social Incentive Fund) and European authorities (URBAN).

The re-arrangement of streets and squares was central in the renewal schemes, including that of Van Wesenbeke Street-Chinatown in 2001. Chinatown's most important thoroughfare now became car-free. Marble lions were installed at the entrance and exit, with 'Chinatown Antwerp' and the names of major Chinese donors engraved on them. This was a compromise as the city government at the time could not install a Chinese arch. The city's refusal here was not prompted by religious or ideological concerns; it had to do with the cables of the tram running on the street. In the meantime the city government gave its permission for the entrance arch, leading a Belgian merchant to file a complaint; the case is still pending. The street lighting was replaced by 18 Chinese lampions in the shape of dragons. Banners with images of China and Chinese characters were also installed, further enhancing the 'Chineseness' of the ethnoscape. For these projects the city sought donations from Chinese associations and individuals. The attempt to make the street car-free, however, was resisted by Chinese merchants; as a compromise the street is not entirely car-free but discourages their use.

Alongside the renewal of streets and squares, the city has improved administrative services in the 'problematic' multicultural area. The 1997 purchase of the once-glamorous but now vacant Permeke Building on De Coninck Square—formerly the Permeke Ford Garage, 'the Mecca of American luxury cars'—led to a new neighbourhood complex housing a council office, a public library and organisations running social projects. The latter include meetings where residents can file complaints as well as assistance to immigrants and their associations for organising cultural and religious events. In this it has partially succeeded. Besides Chinese festivals (Chinese New Year, the birthday of the Buddha, a summer camp for children and the Moon Festival), the social services support multicultural and ethnic events such as the Bazaar Festival featuring a multicultural soccer match, a cultural caravan and an annual African film festival. But despite the urban renewal and the many festivals, problems such as crime ranging from pick pocketing to human trafficking, drug dealing and addiction and the presence of undocumented migrants remain.

Urban renewal and the intervention of public institutions has nevertheless proven indispensable for the neighbourhood's further commodification as an ethnic precinct, thus confirming Kloosterman and Rath's argument on the importance of institutional actors.

Chinatown/Asiatown Brussels: Following the Flow of the Urban Symbolic Economy

This run down and undesirable no-go area of the 1970s has undergone a spectacular spatial make-over: it is now an upbeat, young and trendy neighbourhood. In the 1990s, it began to attract a wide range of stores selling avant-garde designer clothing, mostly Belgian (Ann De Meulemeester, Dries Van Noten, A.F. Vandervorst) but also Japanese (Yohji Yamamoto), footwear (Hatshoe) and glasses (Theo). A flagship store for designers (Annemie Verbeke) followed. In contrast to uptown Brussels, which is more mainstream, bourgeois and mostly French-speaking, the Dansaert area attracts young, avant-garde and mostly Dutch-speaking designers and artists, although it is becoming increasingly diverse.

The rapid transformation was made possible by a mix of trendy restaurants, popular cafés and cultural/educational institutions including a film school and a cultural house for music and dance. The presence of these shops and cultural institutions led to an influx of young creative people and celebrities including rock musicians (Arno), film directors (Dominique Deruddere), theatre makers (Jan Decorte) and food critics. The stores, the institutions, the celebrities, the artists and the aspiring art students all contributed to the neighbourhood's 'creative mood', transforming it into a desirable and upbeat area to work, live, be seen in and meet like-minded people. Shop owners, however, fret over the entry of international labels such as Diesel and Rue Blanche, which threaten the small shops selling unique pieces by local designers—and thus the area's creative 'vibe'.

Contrary to Chinatown Antwerp, the clusters of Asian and Chinese stores in Brussels are adjacent to this trendy part of town. Chinatown Brussels benefits greatly from the area's gentrification, which has diversified the clientele of Chinese restaurants and shops. Trendy Dansaert Street, with its faddish and young image, attracts mostly Belgians and foreign tourists. Chinese shops in the vicinity include Rouge, which sells interesting contemporary Chinese objects, the teahouse Nong Cha and a Thai restaurant (owned by a Chinese from Hong Kong). These establishments cater to the tastes of postmodern urbanites, *bourgeois-bohemiens* as well as domestic and international tourists. The restaurants on Saint Catherine Street also attract a mixed clientele. The Kam Yuen supermarket is patronised by Vietnamese Chinese, Hong Kong Chinese, Mainland Chinese, Taiwanese, Japanese, adopted Belgians of Korean descent, local trendsetters, *bobos*, Flemings, native Brussels inhabitants and EU-migrants living in the city.

Rather than urging architectural intervention to create a Chinatown, Chinese shopkeepers in Brussels are busy selling Chinese/Asian cultural products and services. They are thus unintentionally and indirectly transforming the neighbourhood into a distinctive Asian ethnoscape, especially in the eyes of customers and visitors. The Chinese merchants benefit greatly from their location in trendy Dansaert, especially those who settled there when it was still a run-down area. In this neighbourhood the symbolic economy looms large: Chinese/non-Western products appeal to the fantasy and tastes of the postmodern urbanite through their inter-cultural flavour and texture.

DIVERGING TRAJECTORIES AND COMMON CORE MARKERS

This chapter examined two ethnic precincts in Belgium that at first glance appeared to be relatively similar Asian ethnoscapes. Closer examination, however, revealed divergent trajectories in terms of their historical emergence and local embeddedness, as well as differences in immigrants' socio-economic profiles and neighbourhood's positions within wider urban economies.

The transformation of the Chinese streetscape in Antwerp into a Chinatown was largely made possible by the city's urban renewal policies. While the area developed almost 'organically' into a neighbourhood with a visible presence of Chinese food stores and restaurants, it would have been difficult for the ethnic precinct to redesign itself as a Chinese ethnoscape without the intervention of public institutions. The ethnic commodification of neighbourhoods requires architectural intervention in the physical environment to reshape non-descript areas into themed neighbourhoods. The city government granted official permission for such a Chinese architectural make-over when the entire neighbourhood of Antwerp North, which includes Chinatown, was subject to a general re-arrangement of streets and squares. Drawing on the example of the first North American Chinatowns, the physical 'sinification' process included the installation of marble lions, dragon-shaped lampions and the use of Chinese characters on billboards. In the general imagery of both Chinese and non-Chinese, these symbols are intricately associated with Chinatown. Perhaps in light of their common emigration region, socio-economic background and migration experience—as well as the common Chinese diaspora narrative of suffering and hardship (the practice of 'eating bitterness' or *chi ku*)—the Chinese in Antwerp felt compelled to emulate the first North American Chinatown.

They share the Chinese migration dream, which subscribes to the 'from rags to riches' narrative.

Alongside the theatricalisation through physical intervention, the Chinese mood is reinforced by cultural events that have become increasingly public as well as global. Chinese New Year is celebrated in Chinatowns around the world, while other festivals such as the Moon Festival and the birthday of the Buddha have become public festivities in major cities. Though local city governments work closely with Chinese associations to support these events, their willingness to physically transform ethnic precincts with emblematic cultural markers cannot be taken for granted; most municipalities in Western Europe are reluctant to reshape neighbourhoods into distinctive, 'exotic' ethnoscapes. In the case of Antwerp, the local government's support of the streetscape's sinification was an indirect outcome of its policies to reduce inter-ethnic tensions, foster social cohesion and promote economic mobility among immigrant entrepreneurs in highly mixed neighbourhoods. Immigrants were encouraged to form cultural associations to organise events, for their own groups and, preferably, for all of the area's residents. Given the presence of Chinese businesses and associations in this multicultural precinct, the eagerness of the settled Chinese community to have a Chinatown and the enterprising attitude of some employees at the local social service, the transformation of the streetscape into a Chinatown became a reality.

Chinatown Brussels followed a different trajectory. Its merchants are loosely if at all related to Belgium's main diasporic Chinese communities originating from Hong Kong, Wenzhou or Qingtian. The merchants in Chinatown Brussels consist of two major groups. The first are twice-migrants, Vietnamese or Indonesians of Chinese descent who arrived as refugees. Their links to China can be tangible in the form of language retention or intermarriage; for others the links are imagined. More recent arrivals include immigrants from Mainland China who arrived as students or spouses. Given their differences in background, age and social class, interaction among Chinese merchants is minimal; nor are they inclined to join existing compatriot associations. Empowered by educational attainment and language proficiency, they do not feel compelled to organise themselves into merchant associations. Their entrepreneurial spirit, business experience and personal social networks form the backbone of their success.

Chinese entrepreneurs in Brussels also benefit from their location in a trendy and upbeat neighbourhood, having adapted successfully to urbanite tastes through their manipulation of interiors and choice of products. In the contemporary market place, bustling activity and exotic shabbiness have become curiously compelling and enticing. Despite the absence of a high concentration of exclusively Chinese stores, the neighbourhood exudes an Asian 'mood' while catering to a diverse clientele—Chinese as well as non-Chinese, male and female, of different age groups. One can smell, eat and buy Chinese and other Asian foods and spices. One can get a

Chinese medical treatment, an Asian haircut, or book a flight in an Asian travel agency. Chinese entrepreneurs have seized this opportunity to adapt and expand their businesses, thereby carving a cosmopolitan/ethnic niche within this trendy neighbourhood.

Their differences notwithstanding, both ethnic precincts—and for that matter, most Chinatowns around the world—share some core common markers. Chinatown on display is the place to sample Chinese and other Asian cuisines, to buy exotic products and produce, and to enjoy Chinese festivals. It is a place one visits in groups, either with family or friends. It is accessible as services and goods accommodate the tastes and needs of local customers. It welcomes people from all walks of life with its prices and wide range of products. Last but not least, the mood is noisy, generous, energetic, colourful and happy—it has a high 'feel good' factor. Yet it surpasses the usual theme park as it is a living neighbourhood where Chinese people are engaged in all kinds of activities. It is a fantasy area populated by real people. It is a place for consumption, a place consumed by Chinese and non-Chinese alike.

REFERENCES

Anderson, K. (1990) '"Chinatown re-oriented": A critical analysis of recent redevelopment schemes in a Melbourne and Sydney enclave', *Australian Geographical Studies*, 28: 131–54.

Appadurai, A. (1990) 'Disjuncture and difference in the global cultural economy' *Theory, Culture and Society*, 7: 295–310.

Chow, E.N. (1996) 'From Pennsylvania Avenue to H. Street, NW: The transformation of Washington's Chinatown', in Cary, F.C. (ed.) *Urban Odyssey: A Multicultural History of Washington, D.C.* Washington: Smithsonian Institution Press, 190–207.

Christiansen, F. (2003) *Chinatown, Europe: An Exploration of Overseas Chinese Identity in the 1990s*. New York: Routledge Curzon.

City of Antwerp (2004) *Antwerp Station and Surroundings: An Integral Approach*. Antwerp: Eddy Schevernels.

Collins, J. (2007) 'Ethnic precincts as contradictory tourist spaces', in Rath, J. (ed.) *Tourism, Ethnic Diversity and the City*. London and New York: Routledge, 67–86.

Conforti, J. (1996) 'Ghettos as tourism attractions', *Annals of Tourism Research*, 23(4): 830–42.

Featherstone, M. (1995) *Undoing Culture: Globalization, Postmodernism and Identity*.
Newbury Park: Sage.

Hannerz, U. (1996) *Transnational Connections: Culture, People, Places*. London: Routledge.

Hannigan, J. (1998) *Fantasy City: Pleasures and Profit in the Postmodern Metropolis*. London and New York: Routledge.

Hsu, H.P. (2007) *The Emergence of Chinatown? The Chinese Entrepreneurial Enclave in Brussels*. Unpublished MA thesis, European Studies, Catholic University of Leuven.

Kloosterman, R. and Rath, J. (eds) (2003) *Immigrant Entrepreneurship: Venturing Abroad in the Age of Globalization*. Oxford and New York: Berg.

———. (2001) 'Immigrant entrepreneurs in advanced economies: Mixed embeddedness further explored', *Journal of Ethnic and Migration Studies*, Kloosterman, R. and Rath, J. (eds) *Special Issue on 'Immigrant Entrepreneurship*, 27(2): 189–202.

Kloosterman, R., van der Leun, J. and Rath, J. (1999) 'Mixed embeddedness: (In) formal economic activities and immigrant businesses in the Netherlands', *International Journal of Urban and Regional Research*, 23(2): 253–67.

Logan, J.R., Alba, R. and Jones, B.J. (2003) 'Enclaves and entrepreneurs: Assessing the payoff for immigrants and minorities', *International Migration Review*, 37(2): 344–88.

Pang, C.L. (2007) *EntrepreNoord. Worden multiculturele wijken hip? Anders kijken naar achterstand. Immigrantenondernemerschap en een creatieve benutting van achterstandswijken.* Antwerp: Stad Antwerpen.

———. (2003a) 'Too busy walking, no time for talking: Chinese small entrepreneurs, social mobility and the transfer of cultural identity in Belgium, Britain and the Netherlands', in Harzig, C. and Juteau, D. (eds) *The Social Construction of Diversity: Recasting the Master Narrative of Industrial Nations.* New York and Oxford: Berghahn Books, 62–82.

———. (2003b) 'Belgium: From proletarians to proteans', in Kloosterman, R. and Rath, J. (eds) *Immigrant Entrepreneurs: Venturing Abroad in the Age of Globalization.* Oxford and New York: Berg, 195–212.

Pang, C.L. and Hauquier, G. (2006) *Chinatown Antwerp.* Antwerp: Stad Antwerpen.

Pang, C.L. and Rath, J. (2007) 'The force of regulation in the Land of the Free: The persistence of Chinatown, Washington DC as a symbolic ethnic enclave', in M. Ruef and M. Lounsbury (eds) *The Sociology of Entrepreneurship* (Research in the Sociology of Organizations ,Vol. 25). Amsterdam: Elsevier, 191–218.

Pieke, F. and Benton, G. (1998) *The Chinese in Europe.* New York: St. Martin Press.

Portes, A. and Jensen, L. (1987) 'What's an ethnic enclave? The case for conceptual clarity: A comment', *American Sociological Review*, 52(6): 768–73.

Portes, A. and Manning, R. (1986) 'The immigrant enclave: Theory and empirical example', in Olzk, S. and Nagel, J. (eds) *Competitive Ethnic Relations.* New York: Academic Press, 47–68.

Rath, J. (2007) (ed.) *Tourism, Ethnic Diversity and the City.* London and New York: Routledge.

Rijkschroeff, B. (1998) *Ethnisch ondernemerschap. De Chinese horecasector in Nederland en in de Verenigde Staten van Amerika.* Capelle a/d Ijssel: Labyrint Publication.

Sanders, J. and Nee, V. (1987) 'Limits of ethnic solidarity in the ethnic enclave', American Sociological Review, 52: 745–67.

Shaw, S., Bagwell, S. and Karmowska, J. (2004) 'Ethnoscapes as spectacle: Reimaging multicultural districts as new destinations for leisure and tourism consumption', *Urban Studies*, 41(10): 1983–2000.

Shen, J. (2007) 'America's new Chinatowns: three new techniques of urbanism', *Urban China*, 23: 89–97.

Song, M. (1999) *Helping Out: Children's Labour in Ethnic Businesses.* Philadelphia: Temple University Press.

Taylor, I. (2000) 'European ethnoscapes and urban development: The return of Little Italy in 21st century Manchester', *City*, 4(1): 27–42.

Waldinger, R. (1993) 'The ethnic enclave debate revisited', *International Journal of Urban and Regional Research*, 17(3): 428–36.

Waldinger, R., Aldrich, H., Ward, R. and Associates (1990) *Ethnic Entrepreneurs: Immigrant Business in Industrial Societies.* Newbury Park: Sage.

Watson, J. (1977) 'The Chinese: Hong Kong villagers in the British catering trade', in Watson, J. (ed.) *Between Two Cultures: Migrants and Minorities in Britain*. Oxford: Blackwell Publishers, 181–213.

Yamashita, K. (2003) 'Formation and development of Chinatown in Japan: Chinatowns as tourist spots in Yokohama, Kobe and Nagasaki', *Geographical Review of Japan*, 76: 910–23.

Zhou, M. (1992) *Chinatown: The Socio-economic Potential of an Urban Enclave*. Philadelphia: Temple University Press.

Zukin, S. (1998) 'Urban lifestyles: diversity and standardization in spaces of consumption', *Urban Studies*, 35(5–6), 825–39.

———. (1995) *The Cultures of Cities*. Cambridge: Blackwell.

———. (1991) *Landscapes of Power. From Detroit to Disney World*. Berkeley: University of California Press.

Zukin, S., Baskerville, R., Greenberg, M., Guthreau, C., Halley, J., Halling, M., Lawler, K. (1998) 'Vegas in the urban imagery: Discursive practices of growth and decline', *Urban Affairs Review*, 33(5): 627–54.

4 Kreuzberg's Multi- and Inter-cultural Realities

Are They Assets?[1]

Johannes Novy

INTRODUCTION

Despite its frequently noted 'exceptionalism' (for a discussion see Cochrane and Jonas 1999), Berlin's development after the 1989 fall of the Berlin Wall exemplifies many of the main trends shaping contemporary cities. In line with tendencies commonly associated with the dynamics of urbanisation under contemporary globalised capitalism—variously referred to as post-industrial, postmodern, post-Fordist, or more derogatively, as 'neoliberal urban restructuring'—Berlin's economy has shifted away from manufacturing towards service and knowledge-based industries (Krätke 2004; Cochrane and Jonas 1999; Ward 2004). City politics has become the politics of growth and entrepreneurialism (Mayer 2006) while growing social inequality, poverty and the spatial segregation of poor and migrant populations have led to patterns of polarisation and deprivation previously unknown in Germany's old and new capital (Häußermann and Kapphan 2000). While scholarly attention has focused on these and other related developments such as the physical reconstruction of the 'New Berlin', much less attention has been paid to other aspects of Berlin's transformation from a divided into a united city. One such area is the role played by tourism in the city's restructuring over the last two decades.

The remarkable growth of tourism since 1989—along with such industries as media, arts, music, software and life sciences research—has left its mark on the city (Krajewski 2005). The impact is most evident in Berlin's central areas which include most of its top tourist destinations. But it is also felt in less centrally located areas that lack conventional attractions and were not planned—and until recently not marketed—as tourist zones (Novy and Huning 2009). Berlin's Kreuzberg district, a heavily migrant neighbourhood that has long been targeted for urban renewal, is a case in point. Developments in Kreuzberg support the basic premise of this book that (multi-)ethnic neighbourhoods are attracting growing attention as tourist destinations and places of leisure. Consistent with what has been described in the scholarly literature as the (re-)valorisation of multicultural city spaces (see Rath 2005; Hoffman 2003; Shaw *et al.* 2004), the rise of tourism in Kreuzberg has both resulted from and led to a revalorisation and commodification of cultural and ethnic difference and a transformation of how the neighbourhood is understood, represented and treated by the local government, the tourism industry, and other actors in Berlin's governance and urban development arena.

While recent developments in Kreuzberg appear to follow dominant scholarly interpretations of the processes through which previously marginalised multicultural districts become re-imagined and reconstructed as tourist destinations or places of leisure, they also exemplify the *contextual embeddedness* and frequent *incompleteness* of such processes. In the case of Berlin, the 'marketing of diversity' (Hoffman 2003) is not as commonplace as it is in many other large cities of the Western world. The city's roughly 435,000 foreign residents are only superficially integrated into policies that promote Berlin as an internationally open, cosmopolitan metropolis and destination, while Berlin's urban policies and development context further limit ambitions to draw tourists to disfavoured neighbourhoods (Novy and Huning 2009).

Embedded in a discussion of the development of tourism in Kreuzberg after the fall of the Wall and focusing on the role of local government in promoting tourism beyond the city's core, this chapter explores strategies to market Kreuzberg as a desirable place to visit, play and consume, as well as the limitations and constraints on such efforts. In doing so, this article sheds light on the way the re-imagination and repositioning of multicultural districts as leisure/tourism destinations is mediated by specific local conditions. At the same time, it challenges narratives that overstate the totalising tendencies of neoliberal policy agendas and their implications for (multi-)ethnic neighbourhoods.

KREUZBERG: POVERTY, MIGRANTS AND MULTICULTURAL 'FLAIR'

Located south of Berlin's historic centre, Kreuzberg was formally established as a district in 1920, uniting parts of the former Luisenstadt, the southern Friedrichstadt and the Tempelhofer Vorstadt. Most of its housing stock dates from the turn of the twentieth century when Berlin, booming from industrialisation, had the highest population density of any major city in Europe (Richie 1998: 163). A mainly poor, working class neighbourhood, Kreuzberg was among the most densely inhabited parts of the city, with more than 400,000 residents and at times more than 60,000 residents per square kilometre. Five- to six-story tenement buildings fronting the streets hid rows of back and side buildings occupying entire city blocks. Most apartments were small, overcrowded and lacked adequate heating, light and ventilation; in many of the neighbourhood's notorious Mietskasernen ('rent barracks'), 10 or more flats shared a single toilet.

BERLIN

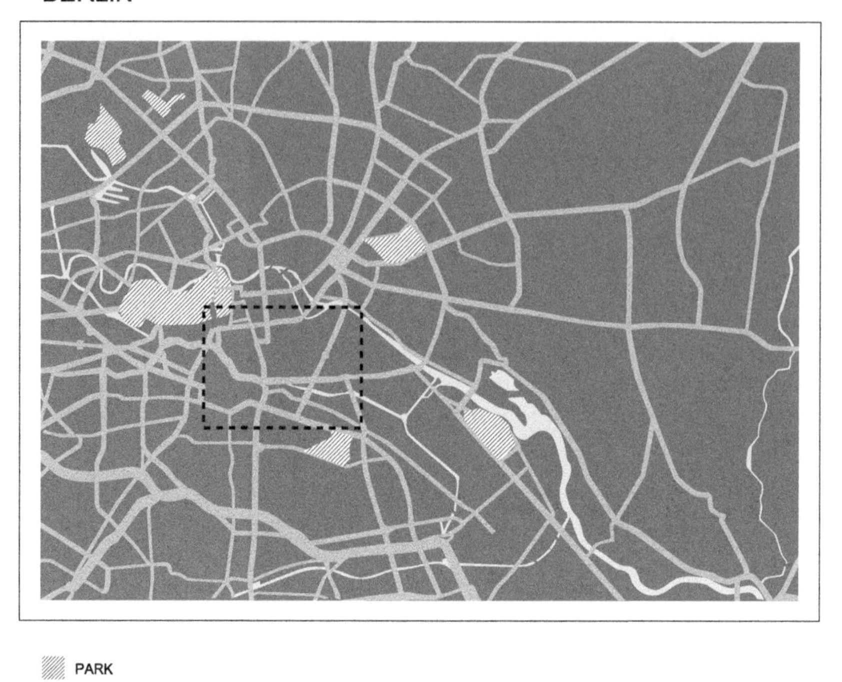

Figure 4.1 Location of Kreuzberg in Berlin. *Map designed by Ana Sala.*

KREUZBERG

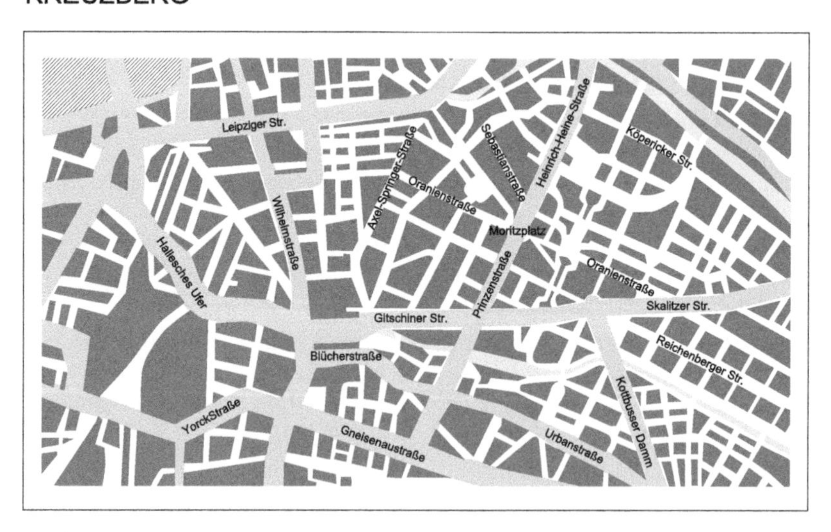

Figure 4.2 Map of Kreuzberg. *Map designed by Ana Sala.*

Much of Kreuzberg was destroyed during World War II, which left less than 60 per cent of its tenement buildings habitable (Diehl *et al.* 2002: 2). The city's subsequent division and the construction of the Berlin Wall in 1961 made the district an isolated area of West Berlin. In 1963 the West Berlin Senate announced its first urban renewal program: 43,000 apartments were to be torn down within 10–15 years, while 24,000 new units were to be built. Located in the immediate vicinity of the Berlin Wall, Kreuzberg was one of the targeted districts. As a result many property owners stopped investing in old buildings while those who could afford to do so moved out of the neighbourhood. Turkish 'guestworkers'—who began to migrate to West Berlin in the thousands in the 1960s—moved with other ethnic minorities and groups at the margins of West German society into Kreuzberg's largely abandoned building stock. In Kil and Silver's words (2006: 96), the district became an 'island of the foreign, the "Other", and the poor' and—due to the high visibility of Turks—known as 'Little Istanbul'.

Over the years, students, artists and other bohemians followed and the neighbourhood became (in-)famous for its mix of alternative lifestyles, multicultural scenes, leftist activism, young art and nightlife (Lang 1998). At the same time, urban renewal continued to threaten the district. Particularly in southeastern Kreuzberg—sometimes named 'SO 36' after the last two digits of its former post code—older housing was cleared to make way for massive housing projects like the Neues Kottbusser Zentrum (NKZ), until this day a symbol of the failures of modernist renewal and the disillusioning consequences of the 'social democratic utopia' of the 1970s. At the same time, the area attracted more and more alternative and countercultural groups—draft-dodgers, punks, squatters, anarchists, left-wing activists—who protested the city's renewal plans and defended the neighbourhood's remaining Wilhelmian building stock. After years of conflict, the orientation of the Berlin state government gradually shifted towards a more 'careful' approach to urban regeneration (Kil and Silver 2006: 97). As part of the International Building Exhibition (IBA) held from 1984 to 1987 in response to criticism of the earlier slum-clearance approaches to urban renewal (Kil and Silver 2006: 97), traffic was limited in parts of Kreuzberg, courtyards were greened, squats were legalised, and houses and facades renovated. Kreuzberg, or at least parts of it, became an internationally recognised model for sensitive urban renewal.

The fall of the Wall returned Kreuzberg to Berlin's geographic centre. Referring to its location midway between the business centres of East and West Berlin, the local media proclaimed the upcoming transformation of the 'alternative mecca' into an 'exclusive yuppie district' (*Berliner Zeitung* in Lang 1998: 172). But even though the *Mythos Kreuzberg*—multi-ethnic, young, alternative, rebellious—remained attractive and the neighbourhood witnessed new investment and developments, gentrification did not occur to the extent anticipated. Instead Kreuzberg retained its image as a place of cultural pluralism and alternative lifestyles on the one hand and utmost marginalisation on the other. The district's socio-economic deprivation worsened in the 1990s as Berlin fell into economic and fiscal crisis and urban boosters realised that their post-1989 aspirations to make Berlin a first-tier 'global city' rivalling London, Paris and Frankfurt would not materialise. Due to the demise of Berlin's industrial base, growing unemployment, stringent welfare cuts and new patterns of socio-economic polarisation—and against the backdrop of decades of flawed integration policies and concentrations of ethnic poverty—a new downward spiral set in and reinforced Kreuzberg's image as one of Berlin's most troubled areas. Compared to other parts of the city, Kreuzberg today has about the lowest average income and highest rates of unemployment and poverty. Migrants—who account for one-third of Kreuzberg's population (not counting naturalised immigrants)—were disproportionately affected by the disappearance of industrial jobs, while growing resignation and tensions among migrant-background youths only aggravated Kreuzberg's status as a district increasingly *excluded* from mainstream society's social and economic relations.[2]

Alternate Discourses, Contradictory Realities

Different discourses shape popular thinking about the Kreuzberg of today (Lang 1998). One perspective stresses the neighbourhood's socio-economic deprivation and sees it as a symbol for everything wrong about (West) Germany's immigration history. According to this perspective, Kreuzberg is on its way to becoming a 'ghetto' (or has already become one) where Turks and other minorities are being isolated and isolate themselves, where social problems such as unemployment, poverty, violence and vandalism dominate, and where outsiders are less and less present. A second perspective sees Kreuzberg as Germany's only truly multicultural community and a casual place of new lifestyles, multicultural creativity, and inter-ethnic innovation and tolerance (Kil and Silver 2006). While it does not necessarily neglect Kreuzberg's ills, the emphasis lies on its qualities. Sometimes, as for example in the widely available tour guide *Berlin for Young People* (Herden Studienreisen Berlin 2006: 129), even Kreuzberg's problems become assets. Referring to Mayor Klaus Wowereit's widely noted remark that Berlin is 'poor but sexy', the travel guide proclaims 'Kreuzberg is (Berlin's) poorest district—so it's gotta be super sexy!' Accordingly, Kreuzberg is alive and well and on the verge of a 'multicultural renaissance' (Kil and Silver 2006) as migrant communities infuse new energy into the area and creative young people who left in the 1990s for districts in former East Berlin rediscover the neighbourhood. As is usually the case with such disparate views, the reality lies somewhere between them. With its roughly 150,000 residents, Kreuzberg consists of different neighbourhoods with different qualities, problems and opportunities. Some are experiencing urban deprivation and unrest whereas others are relatively stable or are, in fact, being revalorised.

Irrespective of its fragmented, geographically dispersed character, the public imagination came to see Kreuzberg as a whole as 'the ghetto' of West Berlin. Kreuzberg's image still dominates popular thinking about ethnic spaces and cultural diversity in Germany (Kil and Silver 2006: 96). While Turks—Germany's largest and most visible minority—today live and work in other parts of Berlin in equal or greater numbers, Kreuzberg's reputation as a 'Turkish ghetto' or 'Little Istanbul' persists as the district retains its importance for the Turkish Diaspora and Turkish cafés, shops, markets and restaurants feature prominently in its streetscape. According to estimates, roughly one-third of Kreuzberg's businesses are run by people with Turkish backgrounds, many employing family members and serving both German and Turkish customers.

Diversity in Kreuzberg is not limited to Germans and Turks, however. The neighbourhood also counts among its residents refugees from Bosnia and Kosovo, Arabs, Kurdish and Lebanese immigrants, ethnic Germans from Eastern Europe, and many others. They have made Kreuzberg their home and often maintain transnational identities, blending the cultures of Berlin and their imaginary homelands and making Kreuzberg not only truly multicultural but a place of inter-cultural exchange and innovation. In Germany, a nation only slowly coming to terms with its reality as an immigration country, where concepts building upon a singular, bounded notion of 'German culture' as opposed to its minority other(s)—such as *Mehrheitsgesellschaft* (majority society) or *Leitkultur* (leading culture)—continue to have widespread support (see Bloomfield 2003), Kreuzberg's multiculturalism and inter-cultural fusions unsurprisingly provoke conflicting images and emotions. Despite its frequent stigmatisation (Mayer 2006) and the bleak everyday reality of many of its inhabitants, it is precisely this vitality and variety that decades of immigration have brought to Kreuzberg—along with its reputation as a counter-culture hotbed—that fuel its image as one the liveliest and most exciting parts of Berlin.

Destination Kreuzberg

Although lacking traditional forms of high cultural consumption such as museums and historic sites as well as an elaborate tourism infrastructure, Kreuzberg emerged as a destination during the period of Berlin's division when it gained international fame as West Berlin's radical and multicultural centre. Tour buses regularly drove past the squats and bazaar-lined streets that became the manifestations of Kreuzberg's 'otherness'. Travel guides and other media paid attention to the district and informed readers about its history and present conditions as well as the pubs, street markets and theatres popular among locals and Berliners (Lang 1998: 139). Tourism thus became a part of Kreuzberg's daily life in the 1970s and 80s as it drew visitors from other parts of Berlin and Germany as well as (to a lesser extent) from abroad. The proliferation of shops, bars, cafés and cultural attractions that came to underpin the neighbourhood's appeal could not have flourished on the disposable income of the district's disproportionately poor residents alone.

After the Wall fell in 1989, several high-profile developments on Kreuzberg's fringes, including Daniel Liebeskind's acclaimed Jewish Museum, attracted new streams of visitors. At the same time, this heightened concerns about the dilution of the neighbourhood's distinctiveness. While the impact of these developments was less cataclysmic than expected, Kreuzberg remained 'in the (self-)portrayal of the city', as Soysal (2006: 43) puts it, 'the locus of hip and diversity', attracting tourists interested in migrant culture and the district's rebellious past as well as consumers of leisure drawn to the array of specialised stores or the prospect of a fun night out.[3]

Local events like the annual Karneval der Kulturen (Carnival of Cultures) have furthered Kreuzberg's profile as a tourist destination (Soysal 2006: 43),[4] while a proliferation of businesses catering exclusively or primarily to tourists has helped consolidate this status. Between 1993 (the first year statistical data are available) and 2006, the number of formally recognised hotels and guesthouses in Kreuzberg tripled from nine to 27. While many, including the recently opened Holiday Inn City Center, chose to locate in Kreuzberg due to its proximity to the city's central business and tourist districts, others have tried to capitalise on the neighbourhood's appeal among tourists with sophisticated, cosmopolitan tastes. Due to the proliferation of lodging within its boundaries, Kreuzberg experienced increases in overnight stays unequalled by other former West Berlin city districts. In 1993 Kreuzberg registered 56,560 overnight guests and 148,099 overnight stays; by 2006, there were 471,762 overnight guests (+734 per cent) and 1,101,182 overnight stays (+643 per cent) (Statistisches Landesamt Berlin-Brandenburg 2007).[5]

The identity of the visitors in Kreuzberg's lodges, the uncounted visitors staying with relatives or friends, and those who make day trips into the district—as well as their behaviour and reasons for coming—are difficult to determine due to lack of data at the city and particularly local level. Evidence suggests that most visitors to Kreuzberg don't fit the negative polemics that have shaped much thinking about tourists in the past. Most visitors illustrate the often-noted shift away from standardised mass tourism towards more individualised, differentiated travel (MacCannell 1999; see also Rojek and Urry 1997). This supports research on tourism that finds increased 'conviviality' (Maitland and Newman 2004; Maitland 2008) among different groups of city users in post-industrial cities, of visitors who appear to share many of the demographic qualities and consumption and lifestyle preferences of workers and residents in the neighbourhood. This applies to consumers from the dominant culture and international tourists whose demographic and lifestyle characteristics often resemble those in the district's more affluent, better-educated quarters, as well as Germans with migrant backgrounds and foreigners with co-nationals living in the district who come for extended family reunions, weddings and cultural festivals, or simply to shop, eat and stroll.

While the influx of German visitors with migrant backgrounds and foreigners visiting co-nationals living in Kreuzberg—who according to business owners are leading patrons of the area's specialised stores and restaurants—has largely been ignored, the district's growing popularity as a destination has not escaped the attention of public officials. At first glance, developments seem to follow the broad consensus in the scholarly literature (in English-language publications at least) on how disfavoured yet colourful urban precincts have come to be understood, represented and treated in light of the new urban growth dynamics associated with tourism, culture, leisure and consumption (see for example Zukin 1995; Hoffman 2003; Shaw *et al.* 2004; Rath 2005; Dávila 2004; Gotham 2005). The consensus posits that major socio-economic changes over the past decades have created a new competitive environment where working class and minority neighbourhoods once considered obsolete in the wake of post-Fordist urban restructuring are now seen to possess cosmopolitan flair, creative energy and/or distinctive heritage and culture. Today they are widely recognised as important symbolic and material assets and repositioned and reconstructed as 'interesting landscapes' for post-Fordist consumption and production (Zukin 1995; see also Rath 2005).

Clearly, the promotion of urban spaces as tourist destinations and places of leisure has influenced Berlin's trajectory since 1989 (see for example Cochrane 2006; Häußermann and Colomb 2003). As will be discussed below, developments in Germany's capital also reveal the contingent and frequently incomplete nature of efforts to develop and promote disfavoured neighbourhoods as destinations. We thus need to examine the governing and institutional arrangements within which tourism and urban development patterns and practices occur—to examine the specifics of urban (policy) change rather than apply some universal template. For although attempts to boost tourism in disfavoured neighbourhoods like Kreuzberg have been the subject of much policy rhetoric, attempts in practice are limited by Berlin's specific urban policy and development context as well as the authorities' struggle to come to terms with a city of multi-ethnic communities in an increasingly diverse Germany (see Bockmeyer 2006).

DEVELOPING AND MARKETING 'NEW' DESTINATIONS IN THE 'NEW BERLIN'

While Berlin's development since 1989 has been disillusioning in most respects, tourism has blossomed over this period. By most accounts, visitor numbers have more than doubled since the early 1990s, catapulting Berlin into the top league of urban destinations worldwide. As Germany's leading destination for leisure and business tourism, the capital currently draws around 140 million visitors annually, placing it third among Europe's most visited destinations, behind only London and Paris (see BTM 2007a). This growth has translated into tourism's continuously rising importance for Berlin's otherwise weak economy: the sector today makes up 7.5 per cent of the city's aggregate income (5 per cent in 2003), generates more than €1 billion in local taxes (€703 million in 2003) and accounts for over 255,000 of the city's 1.5 million full-time jobs (170,000 in 2003) (BTM 2007b). As visitors flocked to the city in record numbers year after year, Berlin's policy-makers saw that tourism not only promised jobs and revenue but could help (re-)define re-united Berlin's identity, (re-)position the city in the German and international inter-urban competition, and revalorise the vast derelict spaces scattered about the city's core (Häußermann and Colomb 2003). Most of the attention thus focused on the city's historic core in Berlin-Mitte and adjacent Potsdamer Platz. But particularly in recent years, tourism's promoters have begun presenting non-mainstream locations like Kreuzberg and gentrifying neighbourhoods in former East Berlin such as Friedrichshain and Prenzlauer Berg as embodiments of the 'New Berlin'—a dynamic, tolerant, diverse, experimental and youthful place where anything goes and trends are set (see Farias 2007; Vivant 2007).

Exemplifying this growing appreciation of 'off mainstream' destinations, Berlin's first 'tourism concept' released by the city government in 2004 highlighted destinations beyond the city's core as well as visitors' desire to experience the 'authentic' and 'original' (Senatsverwaltung für Wirtschaft, Technologien und Frauen 2004). Often part of broader efforts to boost economic growth and revitalise commercial areas, the government began to promote tourism in areas that were previously off the tourist circuit (see Novy and Huning 2009). For example, resources were made available through economic development programmes such as the 'Local Coalitions for the Economy and Labour' (Bezirkliche Bündnisse für Wirtschaft und Arbeit (BBWA)) to encourage Berlin's 12 borough councils to enhance their marketing and invest in tourism-relevant infrastructure, while a city-wide signage system was developed to facilitate finding one's way in the city. Such efforts, however, have generally been small-scale and their effectiveness remains uncertain. While Berlin's city government contributes to the rise of disadvantaged neighbourhoods like Kreuzberg as destinations through various forms of planning within which tourism and tourism-related activities are implicit—for example through land use, cultural and neighbourhood regeneration policies—strategic planning for directing tourism or developing and marketing destinations remains limited.

Exemplifying the evident mismatch between policy rhetoric and practice, Berlin's public sector-led policy-making on tourism is largely uncoordinated and frequently met by opposition. Not all actors in Berlin's fragmented and conflict-prone governmental and administrative system (see Strom 2001: 25) embrace tourism or support the mobilisation of urban spaces as arenas of leisure. Those stakeholders who generally favour tourism do not always work together; nor do they all believe the industry requires interventionist policies or financial support. Berlin's chronic fiscal distress and the complex and contested interplay of bureaucracies, electoral politics and interests that defines its local politics ensure that the development of tourism continues to be driven by the organised interests of the tourism industry, which are unsurprisingly less concerned with tourism as a means of urban (economic) development than with the maximisation of profits.

Berlin Tourismus Marketing GmbH (BTM)

Mainly funded by the subventions of Berlin's large hotels and other large private and semi-private actors like the city's fair and convention centre (Messe Berlin GmbH), the Berlin Tourismus Marketing GmbH (BTM)—which in 1994 replaced Berlin's Public Transportation Office (Verkehrsamt) as the city's official marketing authority—is arguably the most powerful player among the many actors concerned with the development of tourism in Berlin. A public-private partnership, the BTM has a conventional marketing approach that reflects the interests of its donors, most of whom are active in the city's central areas. The BTM thus targets tourists likely to increase its members' revenues, centres its promotional activities on sites and happenings in the city's central areas, and markets places elsewhere only if it sees them as assets for drawing visitors to Berlin. While the increasingly sophisticated and cosmopolitan lifestyles, tastes and amenities demanded by many tourists; the rising attractiveness of minority and former working class neighbourhoods; and 'alternative', 'creative' and 'off' scenes in the form of bars, cafés, shops and clubs are by no means neglected, many district- and neighbourhood-level actors continue to criticise the BTM for not paying more attention to what is on offer outside the city centre. Particularly its references to Berlin's multicultural and inter-cultural scenes are few and far between.

A Negligible Resource in the Emerging Tourism and Leisure Economy?

While the BTM seeks to sharpen Berlin's image as a *weltoffen* (open to the outside world) cosmopolitan metropolis, Berlin's migrant communities and (multi-)ethnic scenes are—in its view—simply not very valuable for promoting Germany's capital as a destination: 'other cities are more multicultural than Berlin' (in Soysal 2006: 40). Ultimately, the BTM's reservation exemplifies the more general difficulty among stakeholders in appreciating the vitality and variety that migrant communities bring to the city. Contrary to cities in the traditional countries of settler immigration such as Australia, Canada or the United States, stakeholders in German cities continue to struggle with the emergence of (multi-)ethnic districts. Whereas some stress migrant communities' creative cultural, economic and social potential, others—pointing to the failure of integration and urban conflict—problematise their existence and seek to relativise their identities in urban spaces where so-called minority cultures (will soon) represent the majority (Bockmeyer 2006). Because of these differences in opinion, as well as the more general ambivalence among Berliners towards issues related to immigration and ethnic diversity, an institutional infrastructure that promotes Berlin's *de facto* multicultural character and emerging inter-cultural realities has—despite the local government's formal endorsement of cultural diversity—not emerged to the extent it has elsewhere. The above-mentioned biases in marketing, the limited capacities of Berlin's public sector and concerns over the appropriateness of promoting disadvantaged neighbourhoods as destinations further limit the incorporation of precincts like Kreuzberg in marketing efforts and similar activities.

Approaches to promoting tourism and leisure differ between Berlin's districts and neighbourhoods.[6] Whereas public, commercial and third-sector actors in parts of the city such as Prenzlauer Berg and Köpenick have recently made concerted efforts to nourish visitor and leisure economies, other parts of the city have neither adapted to the growth dynamics associated with tourism and leisure consumption nor to the growing competition within the city for tourists and the money they spend.

PROMOTING OR PROTECTING KREUZBERG? NEIGHBOURHOOD POLITICS AND TOURISM

The ascendance of neoliberal growth strategies and the global name recognition Kreuzberg enjoys as Berlin's 'stage for displaying diversity and multicultural flavor and color' (Soysal 2006: 44) notwithstanding, marketing strategies deploying the district's aesthetic, cultural and symbolic assets have not become central in Kreuzberg's local decision-making and planning. To be sure, the local borough administration[7] was involved in several tourism-related initiatives as part of its larger strategy to revitalise the local economy and regenerate commercial areas; together with private and third sector actors, it established the local tourism marketing organisation Multi-Kult-Tour e.V. (MKT) and spearheaded the European Union-funded regeneration program 'District Tourism as a Means of Income Generation' to create opportunities for local residents. Most of these efforts—like the joint marketing venture MKT—were short-lived and failed to produce tangible results. In contrast to other Berlin districts and neighbourhoods, Kreuzberg's efforts to capitalise on its strategic advantages and nourish its visitor and leisure economy remain limited. A comprehensive strategy is nowhere in sight.

Why this is the case is difficult to assess, though there are at least two important factors to consider. First, chronic under-funding has made the municipal borough that contains Kreuzberg financially worse off than most others in Berlin, with local authorities hard pressed to fulfil even their core duties (provision of environmental and social services, etc.). Second, at least some segments of the population and many in Kreuzberg's governance and development arena are resentful of tourism for distorting and diluting local culture and identity as well as for promoting gentrification. While such concerns are not unique to Kreuzberg, they are pronounced as the district's development politics are influenced by a powerful local elite made up of (former) community activists and others involved in the upheavals against the city's urban renewal plans of the 1970s and 80s. Many of these former activists have established themselves in local governance—for example as politicians of the district-dominating Green Party—and with a new generation of activists, 'protect' the neighbourhood against developments they deem inappropriate.[8]

Even though neighbourhood politics has adapted to the changing realities of urban policy-making, and 'neighbourhood' interests such as community services, education and housing now often take second place to economic expansion, Kreuzberg has retained at least some of the deeply embedded—and powerfully defended—sense of place it acquired during earlier battles against urban renewal. The district remains particularly sensitive to developments considered inimical to its character and integrity. Furthermore, businesses with a vested interest in developing tourism are mainly small 'mom-and-pop' operations whose proprietors have little time or cash to invest in joint marketing ventures, while associations representing those who might benefit from more visitors—like the neighbourhood's ethnic entrepreneurs—have yet to discover the attendant opportunities. This lack of activity by ethnic business and community associations also sheds light on another significant barrier to the marketing and promotion of Kreuzberg as an inter- or multicultural destination: the majority society's ambivalence towards immigration and ethnic diversity. This ambivalence arguably explains not only why Berlin's ethnic diversity is not marketed more aggressively, but also why ethnic communities have not positioned themselves more assertively in Berlin's tourism and leisure trade.

As the wider economic and politico-institutional structures migrant communities encounter in Germany differ from those in established immigration countries, the opportunity structure for them to integrate, express themselves, succeed economically and assert their interests in the political arena differs as well (Waldinger *et al.* 1990). Despite decades of migrant organising and networking, politics and community affairs in Kreuzberg remain dominated by the 'receiving' society, while outreach efforts to fuel participation and involvement generally fail (Bockmeyer 2006). Even well-intentioned attempts to celebrate and promote diversity such as the Carnival of Cultures or Multi-Kult-Tour e.V. reveal existing power differentials as most key representatives and stakeholders have no migrant background.

CONCLUSION

Mirroring developments in (multi-)ethnic districts in other European and North American cities (see Rath 2005; Hoffman 2003; Shaw *et al.* 2004) and reflecting widespread changes in contemporary patterns of urban development and travel, tourism and leisure consumption have become integral parts of everyday life in Berlin's Kreuzberg district. Both the cause and consequence of the revalorisation and commodification of cultural and ethnic differences in the neighbourhood, Kreuzberg's rising attractiveness as a site for tourism and leisure consumption has helped change the way the neighbourhood is understood, represented and treated in Berlin's governance and urban development arena. However, the marketing of diversity to the extent it has become commonplace in many other cities in the advanced capitalist world—i.e. a real change in policy towards re-imagining and reconstructing Kreuzberg as a leisure commodity—has not occurred.

As part of the broader shift towards neoliberal policy agendas centred on market-oriented growth and competitiveness, efforts have been made to exploit Kreuzberg's manifold aesthetic, cultural and symbolic qualities through image and marketing strategies. These efforts, however, typically do not involve established institutional structures and are sporadic and limited in scope. Berlin and Kreuzberg's specific governance and development contexts, the continued ambivalence towards Berlin's multi- and inter-cultural realities, and migrants' lack of involvement as legitimate stakeholders in community affairs all inhibit ambitions to draw tourists and leisure consumers to the disadvantaged neighbourhood. Whereas (multi-)ethnic neighbourhoods in North America and other European cities—most notably in Great Britain (Shaw *et al.* 2004) but also in continental Europe, for example in the Netherlands (Rath 2005; van der Horst 2003)—have elaborate strategies to promote tourism emphasising consumption, strategic image management and competitive niche thinking, developments in Kreuzberg have been less cataclysmic and demonstrate the *contextual embeddedness* and frequent *limitations* of efforts to market and develop (multi-)ethnic urban spaces as destinations for tourism and leisure consumption.

How are Kreuzberg's residents and businesses faring under the current situation? Are efforts to strengthen the local visitor and leisure economy aiding or hurting their well-being? Some scholars argue that tourism and leisure revalorise marginalised neighbourhoods at the expense of long-term residents and businesses as most of the benefits are reaped by local elites and community outsiders, while commercial and residential gentrification, rising living costs and other adverse effects place additional burdens on the most vulnerable members of the community (see Zukin 1995; Gotham 2005; Dávila 2004; for discussion see also Shaw *et al.* 2004). At the same time, numerous studies suggest that the greater inclusion of marginalised neighbourhoods in cities' visitor and leisure economies can deliver a range of benefits to those living, working or doing business in them—by generating jobs and revenue, enhancing the quality of neighbourhoods' built environments and infrastructure, strengthening local communities' cultural and civic spheres and counteracting their isolation and stigmatisation (see for example Hoffman 2003; Rath 2005). But for these benefits to materialise, the institutional and regulatory arrangements framing marginalised neighbourhoods are crucial (see for example Rath 2005; Pang and Rath 2007; Fainstein and Powers 2006). Whether Kreuzberg—the people living, working and doing business there—would benefit from greater incorporation into Berlin's bourgeoning tourism and leisure economy thus not only depends on the kind, extent and objectives of future tourism and regeneration policy but on various other factors. These include (but are not limited to): the future development of Berlin's economy and property market, the opportunities provided for residents with non-German backgrounds in general and ethnic entrepreneurs in particular, and the political will to challenge current development strategies that prioritise market-oriented growth at the expense of social goals (see Novy and Huning 2009).

NOTES

1. This paper builds and expands upon the paper 'New Tourism (Areas) in the "New Berlin"', co-authored with Dr. Sandra Huning and published in Maitland, R. and Newman, P. (eds) *World Tourism Cities: Developing Tourism Off the Beaten Track* (Routledge 2008). I would like to thank Sandra Huning and the publisher for allowing me to use passages of the paper for this chapter.
2. While Berlin's fiscal and economic crisis, the dismantling of welfarist-Keynesian and social collectivist institutions, and neoliberal policy agendas all contributed to Kreuzberg's socio-economic malaise in the 1990s, the city-state's public sector has tried to address the attendant problems. Programmes and initiatives to counteract widening socio-spatial divisions are in place in numerous Berlin neighbourhoods, most notably the area-based federal/state neighbourhood regeneration initiative 'Districts with Special Development Needs—the Socially Integrative City' (Soziale Stadt or Social City, for short). But in light of Kreuzberg's manifold structural disadvantages and the cuts in welfare provisions, these efforts are widely seen as drops in the bucket, 'offering a bandaid instead of viable solutions' (Silver 2006: 39; Marcuse 2006).
3. Visitors' interests of course overlap, and if Urry (2002: 1) is correct in arguing that tourism has always been about 'gaz[ing] upon . . . a set of scenes . . . which are out of the ordinary', it seems reasonable to assume that voyeuristic curiosity remains a motive for even the most educated or sensitive visitors.
4. Currently in its 13th year, Germany's largest multicultural festival gathers community and minority groups, cultural initiatives, youth centres and other non-profit associations to celebrate Berlin's diversity. The four-day festival has become a fixture in Berlin's calendar and attracts locals and visitors alike, whom the organisers ironically describe as 'Problemkiezbewohner' and 'No-go-area-goers'.
5. Over the same period (1993 to 2006), the overall number of lodges in Berlin increased from 435 to 578 (+33 per cent), overnight guests from 3,040,466 to 7,077,275 (+133 per cent) and overnight stays from 7,455,151 to 15,910,372 (+113 per cent).

6. Like London and Paris, Berlin has two directly elected tiers of government within the city and is currently split into 12 semi-autonomous boroughs with populations averaging 300,000. While Berlin's boroughs do not raise their own taxes (their financing is derived fully from the city's first-tier authority), they nonetheless have important powers and duties and enjoy significant formal and informal leverage. Each Bezirk is composed of several neighbourhoods, or *Kieze* (a Berlin term referring to 'neighbourhood'), typically marked by a strong sense of community and various development-relevant collective and individual actors.
7. Kreuzberg used to be one of Berlin's municipal boroughs but was merged with adjacent Friedrichshain in a 2001 administrative reform that reduced the number of city boroughs from 23 to the current 12.
8. This was again evident in the recent clashes between McDonalds and local residents/stakeholders who fought the global burger chain's plans to open its first restaurant in the district. While McDonalds was ultimately able to open its eatery in September 2007, protests attracted global media attention and underscored Kreuzberg's image as a stronghold of left-wing, alternative and anti-capitalist sentiment.

REFERENCES

Bloomfield, J. (2003) '"Made in Berlin": Multicultural conceptual confusion and intercultural reality', *International Journal of Cultural Policy*, 9(2): 167–83.

Bockmeyer, J. (2006) '"Social cities" and social exclusion: Assessing the role of Turkish residents in building the "New Berlin"', *German Politics and Society*, 24(4): 49–78.

BTM (2007a) *Destination Berlin.* <http://www.berlin-tourist-information.de/deutsch/presse/download-basistexte/d_pr_basistext_destination-berlin.pdf> (accessed 6 June 2007).

———. (2007b) *Berlin-Tourismus lässt die Kassen klingeln.* <http://www.btm.de/deutsch/presse/download/d_pr_552_wirtschaft.pdf > (accessed 6 December 2007).

Cochrane, A. (2006) '(Anglo)phoning home from Berlin: A response to Alan Latham', *European Urban and Regional Studies*, 13(4): 371–76.

Cochrane, A. and Jonas, A. (1999) 'Re-imagining Berlin: World city, national capital or ordinary place? ', *European Urban and Regional Studies*, 6(2): 145–64.

Dávila, A. (2004) 'Empowered culture? New York City's Empowerment Zone and the selling of El Barrio', *The Annals of the American Academy of Political and Social Science*, 594(1): 49–64.

Diehl, V.S., Sundermeier, J. and Labisch, W. (2002) *Kreuzbergbuch.* Berlin: Verbrecher Verlag.

Fainstein, S. and Powers, J. (2006) 'Tourism and New York's ethnic diversity: an underutilized resource?', in Rath, J. (ed.) *Tourism, Ethnic Diversity and the City.* London and New York: Routledge: 143–64.

Farias, I. (2007) *Touring Berlin.* PhD dissertation, Institute of European Ethnology, Humboldt University of Berlin.

Gotham, K.F. (2005) 'Tourism gentrification: The case of New Orleans' Vieux Carre (French Quarter)', *Urban Studies*, 42(7): 1099–121.

Häußermann, H. and Colomb, C. (2003) 'The New Berlin: Marketing the city of dreams', in Hoffman, L., Fainstein, S. and Judd, D. (eds) *Cities and Visitors: Regulating People, Markets, and City Space.* Oxford and Cambridge: Blackwell, 200–18.

Häußermann, H. and Kapphan, A. (2000) *Berlin—Von Der Geteilten Zur Gespaltenen Stadt? Sozialräumlicher Wandel seit 1990*. Opladen: Leske/Budrich.

Herden Studienreisen Berlin (2006) *Berlin for Young People (English Edition)*. Berlin: Herden Studienreisen.

Hoffman, L. (2003) 'The marketing of diversity in the inner city: Tourism and regulation in Harlem', *International Journal of Urban and Regional Research*, 27(2): 286–99.

Horst, H. van der (2003) 'Multicultural theming: Pacifying essentialising and revanchist effects', in Reisenleiter, M. and Ingram, S. (eds) *Placing History: Themed Environments, Urban Consumption and the Public Entertainment Sphere*. Wenen: Turia+Kant, 175–200.

Kil, W. and Silver, H. (2006) 'From Kreuzberg to Marzahn: New migrant communities in Berlin', *German Politics and Society*, 24(4): 95–121.

Krajewski, C. (2005) 'Städtetourismus im "Neuen Berlin"', in Landgrebe, S. and Schnell, P. (eds) *Städtetourismus*. München and Wien: Oldenbourg Wissenschaftsverlag, 279–95.

Krätke, S. (2004) 'City of talents? Berlin's regional economy, socio-spatial fabric and "worst practice" urban governance', *International Journal of Urban and Regional Research*, 28(3): 511–29.

Lang, B. (1998) *Mythos Kreuzberg. Ethnographie eines Stadtteils 1961–1995*. Frankfurt and New York: Campus.

MacCannell, D. (1999) *The Tourist: A New Theory of the Leisure Class (Third Edition)*. Berkeley: University of California Press.

Maitland, R. (2008) 'Conviviality and everyday life: The appeal of new areas of London for visitors', *International Journal of Tourism Research*, 10(1): 15–25.

Maitland, R. and Newman, P. (2004) 'Developing metropolitan tourism on the fringe of Central London', *International Journal of Tourism Research*, 6: 339–48.

Marcuse, P. (2006) 'The down side dangers in the Social City program', *German Politics and Society*, 24(4): 122–30.

Mayer, M. (2006) 'New lines of division in the New Berlin', in Ulfers, F., Lenz, G. and Dallmann, A. (eds) *Towards a New Metropolitanism: Reconstituting Public Culture, Urban Citizenship, and the Multicultural Imaginary in New York City and Berlin*. Heidelberg: Universitätsverlag Winter, 171–84.

Novy, J. and Huning, S. (2009) 'New tourism (areas) in the "New Berlin"', in Maitland, R. and Newman, P. (eds) *World Tourism Cities: Developing Tourism Off the Beaten Track*. London and New York: Routledge, 87–108.

Pang, C.L. and Rath, J. (2007) 'The force of regulation in the Land of the Free: The persistence of Chinatown, Washington D.C. as a symbolic ethnic enclave', in M. Ruef and M. Lounsbury (eds) *The Sociology of Entrepreneurship* (Research in the Sociology of Organizations, vol. 25). New York: Elsevier, 195–220.

Rath, J. (2005) 'Feeding the festive city: Immigrant entrepreneurs and tourist industry', in Guild, E. and van Selm, J. (eds) *International Migration and Security: Opportunities and Challenges*. London and New York: Routledge, 238–53.

Richie, A. (1998) *Faust's Metropolis: A History of Berlin*. New York: Carroll/Graf.

Rojek, C. and Urry, J. (1997) 'Transformations of travel and theory', in Rojek, C. and Urry, J. (eds) *Touring Cultures: Transformations of Travel and Theory*. London and New York: Routledge, 1–22.

Senatsverwaltung für Wirtschaft, Technologien und Frauen (2004) *Tourismuskonzept für die Hauptstadtregion Berlin*. <http://www.berlin.de/imperia/md/content/senatsverwaltungen/senwaf/publikationen/tourismuskonzept.pdf> (accessed 10 January 2008).

Shaw, S., Bagwell, S. and Karmowska, J. (2004) 'Ethnoscapes as spectacle: Reimaging multicultural districts as new destinations for leisure and tourism consumption', *Urban Studies*, 41(10): 1983–2000.

Silver, H. (2006) 'Social integration in the New Berlin', *German Politics and Society*, 24(4): 1–48.
Soysal, L. (2006) *World City Berlin and the Spectacles of Identity: Public Events, Immigrants and the Politics of Performance*. Migration Research Program, Koç University.
Statistisches Landesamt Berlin-Brandenburg (2007) Commissioned Data Set.
Strom, E. (2001) *Building the New Berlin: The Politics of Urban Development in Germany's Capital City*. Lanham: Lexington Books.
Urry, J. (2002) *The Tourist Gaze*. London: Sage Publications.
Vivant, E. (2007) 'Towards a non-human anthropology of tourism'. Paper at the meeting of the Association of Social Anthropologists, London, April 2007.
Waldinger, R., Aldrich, R., Ward, R. and Associates (1990) *Ethnic Entrepreneurs. Immigrant Business in Industrial Societies*. Newbury Park: Sage.
Ward, J. (2004) 'Berlin, the virtual global city', *Journal of Visual Culture*, 3: 239–56.
Zukin, S. (1995) *The Cultures of Cities*. Malden: Blackwell.

5 Sanitising the Metropolis of Leisurely Consumption

A Missed Chance to Re-invent Entrepreneurial Dynamism in Sulukule, Istanbul

Volkan Aytar and Süheyla Kırca-Schroeder

Over the last decade or so, the national and local governments together with corporate groups have sought to reposition Istanbul as a globalising metropolis of finance capital, services, tourism, entertainment and consumption. However, its accompanying neoliberal, top-down 'urban transformation' (*Kentsel Dönüşüm* or KD)[1] policies have excluded a local 'ethnic' population—the Roma of Sulukule—from the process. This community and neighbourhood could otherwise have capitalised on its human and social capital to re-invent its long-standing history of informal entrepreneurial dynamism, an opportunity that is now being squandered.

While consumers and members of the 'critical infrastructure' have promoted selective 'ethnicisation' in the entertainment and tourism sectors, this did not help to revitalise the ethnic neighbourhood and its leisure economy. The neighbourhood and its entrepreneurs were instead seen as 'obstacles' to be cleared, to make way for new sites of residence, leisure, tourism and entertainment.

The story of the Roma of Sulukule illustrates the interaction of consumers, producers, the critical infrastructure and government regulation in a way that informs theoretical debates on urban governance, political economy and culture, the revalorisation of urban space and ethnic entrepreneurship. The chapter first provides a socio-historical account of the neighbourhood, emphasising the informal role of the Roma in Istanbul's entertainment and consumption economy. It then reviews earlier as well as more recent KD policies in Istanbul, focusing on the neoliberal turn in urban governance and its impact on urban spaces, local economies and the leisure sector. We then examine how these transformations have impacted upon Sulukule, its neighbourhood and community organisations, and their relations with the agents of urban governance and NGOs. The final section draws some theoretical and comparative conclusions.

SULUKULE AND ROMA'S ROLE IN THE ENTERTAINMENT WORLD

Sulukule, situated on the historic peninsula (see map), is one of the oldest Roma-populated neighbourhoods in Istanbul. Some scholars trace its history back to the eleventh- or twelfth-century Byzantine Empire.[2] The neighbourhood's population increased after the Ottoman conquest in 1453, when Roma groups engaged in basketry, metalwork and horse-raising were sent there to revive the local economy (Yılgür 2007). Forced or voluntary population movements such as these were an important part of the empire's demographic policies whereby different groups were moved to various regions and/or neighbourhoods to 'balance' one another (see İnalcık 1997).

The Roma's place, then, should be seen within the larger context of imperial demographic policies and social economy, organised along a functional 'ethnic division of labour' whereby different communities were channelled to particular vocations and localities. Note, however, that the use of the term 'ethnic' is problematic for Ottoman times, where confessional divisions were based on the *millet* system.[3] Although the system did not entail the forced vocational, sectoral or employment-based clustering of *millets* or other confessional/'ethnic' groups, there remained certain

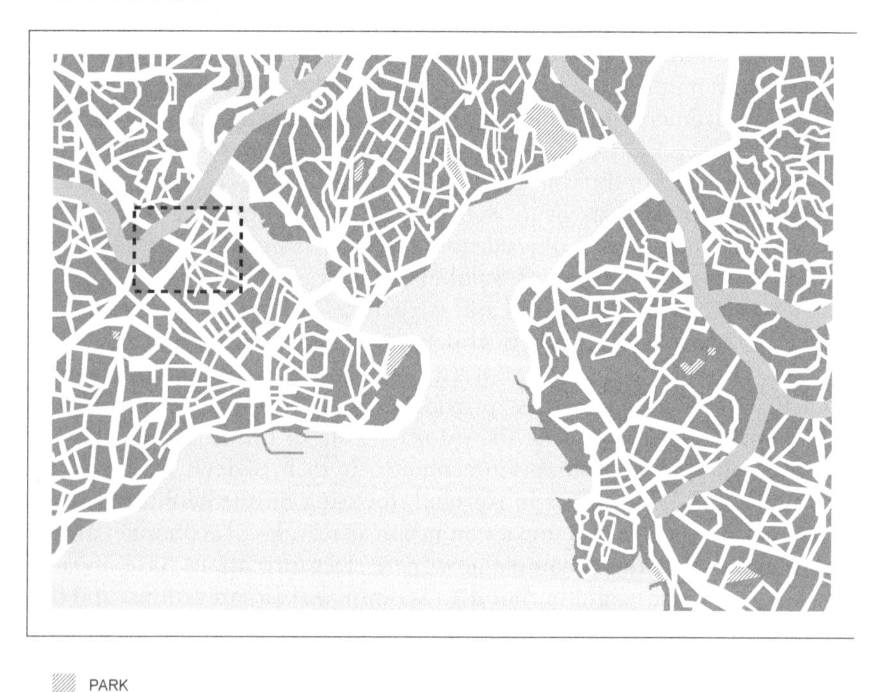

Figure 5.1 Location of Sulukule in Istanbul. *Map designed by Ana Sala.*

economic 'expectations'. The Roma as a Muslim, non-*millet* group were mainly involved in basketry, metalwork, horse-raising, entertainment and related services. As will be discussed below, the Roma's 'inclination' towards consumption-oriented services should be seen in relation to larger socio-economic forces and administrative arrangements.

The Roma of Sulukule have been prominent in the world of entertainment since the seventeenth century as musicians, dancers, fortune-tellers, acrobats and illusionists. According to Evliya Çelebi (Kayaoğlu, 1996), 'Entertainment Patrols'—mobile 'troops' of entertainers, musicians, dancers and other performers dispatched to social occasions, celebrations, weddings, etc.—were dominated by Roma. Akçura (2007) further claims that these patrols were the historical basis of Sulukule's 'entertainment houses' (*Eğlence Evleri*). Public celebrations, gatherings and feasts as well as more private spaces of entertainment and consumption were dominated by Roma performers, dancers and musicians.[4]

Roma personalities served in the Ottoman court and within elite families as music instructors, composers and performers—some even becoming famous in the early recording industry[5]—while Roma musicians, performers, dancers, fortune-tellers and others remained active in popular circles, especially at weddings and feasts.[6] Muslim Roma had a 'parallel', complementary position to Christian Greeks and Armenians widely involved in the restaurant, *meyhane* (winehouse) and other drinking/entertainment sectors, where it was usually 'inappropriate' for Muslim groups to be active.

SULUKULE'S ENTERTAINMENT HOUSES: LOCI OF A PERIPHERAL LEISURE ECONOMY

All accounts point to the vital yet highly informal role the Roma played in the empire's entertainment and consumption economies. This role—which seems to have continued after the foundation of the Turkish Republic in 1923—had a clear spatial dynamic which explains the importance of Sulukule as an ethnically identifiable neighbourhood constituting the loci of the entertainment economy. Due to discrimination, negative stereotyping and frequent criminalisation, Sulukuleans have a tight, solidarity-based community network (see Kırca-Schroeder and Somersan 2007). During Kırca-Schroeder's fieldwork in 2006 and 2007, numerous musicians, performers and entrepreneurs underlined these negative attitudes towards themselves, and likened their situation to that of African Americans.[7]

Sulukule was generally marginalised and suspect in the eyes of mainstream society. Although at times riddled with internal divisions and conflicts, it kept its character as an ethnic neighbourhood surrounded by non-Roma groups including Turks, Greeks and later, Kurds and others.

Roma had an especially strained relationship with Kurds, some of whom were displaced persons from conflict-ridden south-eastern Anatolia. Inter-ethnic tensions and marginalisation by mainstream society contributed to the strength of community solidarity in the neighbourhood.

One of Sulukule's most important spatial markings was established through its tradition of managing 'entertainment houses' or informal 'listen and drink' establishments (Akçura 2007). In these establishments one could rent the entire house, hall or room for an evening and night, entertained by musicians and belly dancers while being served food and alcohol. Operated by family heads with members of immediate or connected families working in different performing and catering roles, these small, family-based enterprises were the loci of Sulukule's entertainment economy. 'Entertainers' included male and female musicians and performers,[8] while other 'employees' were involved in cooking and catering. These businesses were usually also the residences of the entrepreneurs and employees.

In one of the first few written accounts, Beler (1946) portrays the lively atmosphere in these establishments where women played and danced in clean, richly decorated rooms. He notes that customers often came from distant districts and required reservations well in advance. Though highly 'informal'—none of the houses had official permits or liquor licenses, while none of the 'employees' were formally employed or registered—entertainment houses were popular and served important social functions both for their clienteles and their 'entrepreneurs' and 'employees'.

These functions can be seen from two distinct yet related angles. First, entertainment houses were among the alternatives to the spaces for 'appreciating (high) cultural taste' shaped by the cultural policies of Kemalist modernisation, which either repressed or controlled popular forms of music and entertainment. Many of the latter only survived informally.[9] In this perspective, entertainment houses were important informal spaces of cultural production and consumption. Especially during the republican period when the non-Muslim population declined drastically, Roma moved into the entertainment sector to cater to this social need.

Second, entertainment houses can be seen as 'safe havens' for mostly male customers living in a conservative society that controlled the consumption of alcohol, listening to non-religious music and watching female dancers. During fieldwork, some former entertainment house owners, musicians and performers claimed that their type of entertainment originated in Konya and Karaman, two central Anatolian provinces known for their religious orthodoxy and communitarian repression, and that their migrant forefathers brought these practices to Istanbul's Sulukule.

Owners and musicians/performers argued that such practices were known as *oturak âlemi*, informal at-home gatherings of men listening to music, watching female dancers and consuming alcohol.[10] Such gatherings reportedly still take place all around central Anatolia and elsewhere, usually in conservative smaller towns and cities. In this sense, entertainment houses can be seen as the continuation of an informal—perhaps even *carnivalesque*[11]—Anatolian tradition of social occasions/performances taking place under the watchful eyes of cultural conservatism. In this perspective, the Roma of Sulukule are informal entrepreneurs shouldering important social functions, gaining economically but becoming targets of a moralistic and criminalising gaze at the same time.[12]

Entertainment houses had numerous other functions for their clienteles and caterers. Well-known singers, artists and celebrities were reportedly among the regulars, increasing the attraction of the houses especially for members of the 'critical infrastructure', newspaper and magazine columnists, and national and international tourists.[13] Zeki Müren, Müzeyyen Senar and other celebrity musicians reportedly visited the establishments not only for leisure, but also to recruit accompanying musicians and to enrich their own repertoires. In this sense, the houses were key centres of cultural production and consumption.

Entertainment houses also operated as informal schools for numerous Roma musicians who later gained nation-wide and even international prominence and fame.[14] Sulukule musicians were often invited to wealthy family homes to play at private celebrations and parties and were also hired as accompanying musicians at drinking establishments as well as at recording sessions, generating economic gains for the residents and middlemen who helped arrange such appearances. Nor were the social and economic functions of the houses limited to providing jobs for their immediate entrepreneurs and employees. The entertainment houses created secondary circles of employment around them, including tobacco and spirits shops as well as neighbourhood taxis dispatched to shuttle clients from distant neighbourhoods.

DIVERSE PERCEPTIONS OF COMMUNITY SOLIDARITY

The 'Golden Age' of Sulukule's entertainment houses was between the 1950s and 1970s. In the early 1990s, there were 34 establishments providing jobs and livelihoods to more than 3,500 people (it is unclear whether this included only those who were immediately employed, or also those who gained indirectly). Following the demolition of all entertainment houses in the mid-1990s, an estimated 30–35 musicians remained employed in entertainment/dining establishments (all outside of Sulukule) or found occasional employment within the recording industry or elsewhere (see Kalafat and Aydın 2007).

Fieldwork revealed that most residents, entrepreneurs and employees regarded the entertainment houses positively, though they also acknowledged that the supposed notoriety of the establishments made their neighbourhood an 'easy target'. Entrepreneurs and employees angrily refuted allegations of sex trade and other illicit activities, and claimed such allegations were based on widespread anti-Roma sentiments. Some mothers and fathers of female dancers claimed that they always waited outside to escort their daughters after performances ended.

One elderly musician described the neighbourhood as follows: 'Every night it was like the Rio carnival. There were many customers here, dancing and enjoying themselves'. An elderly woman remembered the 1950s: 'in the good old days, music and dance would kick off in the late afternoon and go on until the next morning. Young and old were out in the street; braziers were put in front of houses. Incense was lit to avert the evil eye and to bring good luck to the neighbourhood. Our green flag, which contained many yellow stars, would hang on our walls. We would play and dance all night'. Such celebratory accounts suggest that community identity and feelings of solidarity were strong in the neighbourhood.

Interestingly enough, some residents and musicians had a much more negative view of entertainment houses and claimed they 'degenerated' over time, making the entire neighbourhood 'suspect'. Some residents and musicians claimed that in some houses the sex trade and other illicit activities were indeed going on, and that 'outsiders'—i.e. 'other' 'nomadic/migrant' (*göçer*) Roma people, including those from other provinces such as Adana—were to blame.

This claim is important as it points to interesting dynamics within neighbourhood-level community identification. While some residents proudly and unapologetically call themselves either Roma ('*Roman*') or Gypsy ('*Çingene*'), others insist on drawing a clear distinction between the resident, 'native' *Roman/Çingene* of Sulukule and the 'outsiders', the migrant and/or nomadic *Çingene* ('*göçer çingene*') 'naturally' prone to illicit activities who make their otherwise 'clean' neighbourhood suspect. Interestingly, some respondents within this second group called themselves '*Roman*' and the outsiders '*çingene*'. A third group, reportedly with more explicit Islamic self-identification, refused to be called either *Roman* or *Çingene* and claimed themselves to be 'real Turks'. Indeed, as will be discussed later, conservative residents were happy to see the entertainment houses destroyed and generally supported the KD plans.

Diverse foundational myths imply that identity remains a contested terrain, with some residents resorting to a narrative of 'otherness' directed at the 'outsiders'. Such 'othering' mechanisms were easily observable especially in narratives about Kurds. Even some of those residents who proudly called themselves *Roman* or *Çingene* carefully underlined the fact that 'they were not separatists' and took pride in being Turks. Most residents acknowledged that they had rather strained relations with Kurds and that they did not like their 'violent' ways. These diverse perceptions can perhaps be best understood within the context of Istanbul's urban transformation since the 1950s, and its impact on the Sulukule community, its entertainment houses and social economy.

URBAN TRANSFORMATION, URBAN GOVERNANCE AND SULUKULE

Since the 1950s, Sulukule, along with numerous other neighbourhoods and districts, has been subject to waves of 'urban transformation' and 'renewal'. These schemes reshaped Sulukule, its community, entertainment houses and entrepreneurial dynamics in important ways. In 1957, the market-friendly Democratic Party (DP) government initiated '*Hausmannesque*' urban projects that, in addition to other projects all around Istanbul, razed numerous Roma neighbourhoods south of Sulukule to pave way for large boulevards. The demolition also hit parts of Sulukule.[15] What remained of the neighbourhood was formally merged with the nearby 'Sultan[a]' neighbourhoods in Fatih district, leaving 'Sulukule' as a denotation of popular parlance rather than an administrative unit.

Interestingly enough, while parts of the neighbourhood were demolished by the government and numerous Roma residents had to relocate to

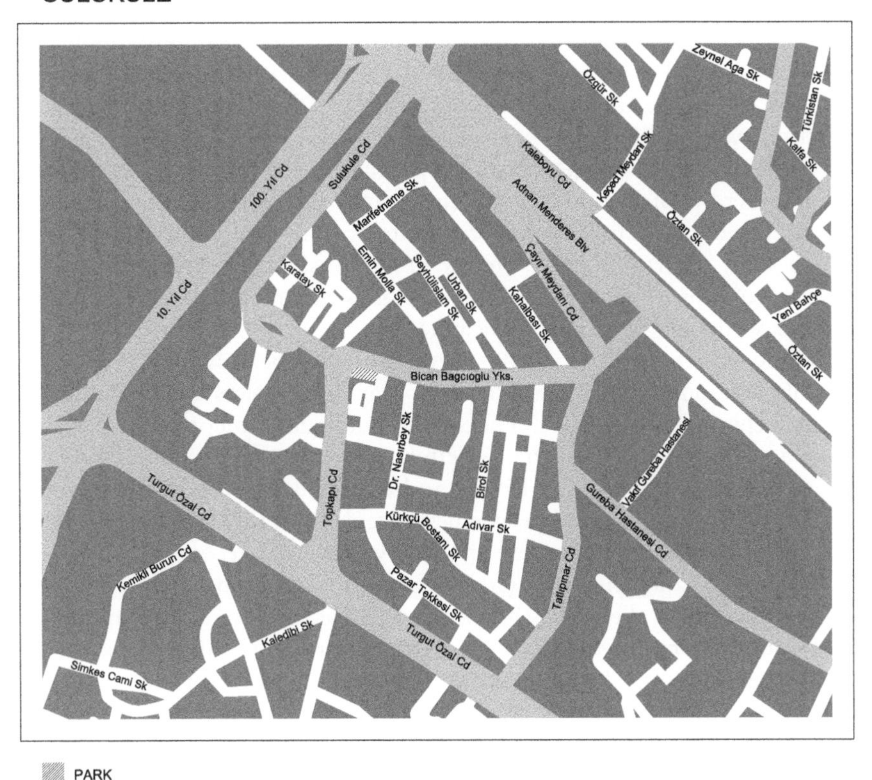

Figure 5.2 Mao of Sulukule. *Map designed by Ana Sala.*

other Istanbul neighbourhoods, the DP government had a short-lived—and highly rhetorical—project to replace the informal entertainment houses with state-initiated and regulated 'Performance Centres' catering to the emerging tourist industry (Akçura 2007). There were plans for 'Sports and Cultural Complexes' as well. This could perhaps be seen as an early example of the national and local governments' interest in capitalising on Sulukule's tourism potential.

Official interest in Sulukule's tourism potential continued after the 1960 military coup. In 1969, the Ministry of Tourism supported the establishment of the 'Sulukule Tourism Renewal and Protection Association' which organised the 'International Gypsy Festival' with the participation of numerous Roma musicians from the Balkans and elsewhere in Europe. While the association mainly strove to make Roma culture and music more visible, one of its underlying aims was to help re-establish Sulukule in its pre-1957 social meaning and existence as an administrative/municipal unit. When these demands were not met by the local municipality, the association was voluntarily disbanded in 1971. The association was nevertheless a significant step in neighbourhood organising, particularly for community-led self-help initiatives.

In 1985, community organisers and entrepreneurs in Sulukule applied to the Ministry of Tourism with a project to establish 'Performance Centres'—reminiscent of the DP's then defunct project. While the Ministry's initial reaction was positive, the local government did not allow the project to proceed. Still, it gave impetus to the opening of new entertainment houses, the number of which rose to 34 by the early 1990s, thereby contributing to the livelihoods of many Sulukule residents (Kalafat and Aydın 2007). Sulukule's local entertainment economy grew considerably over these years, helping to improve the neighbourhood's physical and architectural appearance as well as its store of social capital.

While earlier waves of urban renewal had significantly impacted on the neighbourhood and its social economy, they nevertheless kept community-led entrepreneurialism alive. They even contributed to community organising and helped shape efforts to better utilise the neighbourhood's social capital. But since the early 1990s, and more importantly since the mid-2000s, urban transformation schemes have dealt deadly blows to the Sulukule neighbourhood, its Roma community and leisure economy.

The surveillance and criminalisation of Sulukule and its entertainment houses peaked in the early 1990s when Saadettin Tantan was appointed vice-commissioner of the Istanbul police department. Tantan was known for his toughness in 'cleaning up' socially 'un-hygienic' locations and establishments. Under his orders, Fatih district's notorious police chief, Süleyman Ulusoy, launched a ferocious attack on the entertainment houses.[16] In Sulukule, virtually all entertainment houses had to close under violent pressure from the police, a development that took a huge toll on the neighbourhood. When Tantan was elected mayor of Fatih district in 1994, the remaining few entertainment houses were shut down, pushing most residents into helplessness and poverty. Since the mid-1990s, Sulukule has been an impoverished neighbourhood with unemployment rates far higher than the average in Istanbul and Turkey (Foggo 2007).

These and subsequent developments in Sulukule need to be seen within the broader context of neoliberal urban transformation in Istanbul and Turkey. Following the 1980 coup, Turkey shifted from state-led import-substituting industrialisation to a market-oriented, export-promoting growth strategy. In this process, Istanbul emerged as an aspiring 'global city'[17] promoting itself as a metropolis with world-class facilities such as offices in skyscrapers, shopping and convention centres, hotels, cafés and restaurants, but also with 'local' characteristics that make it an 'exotic' choice in an apparently endless sea of options for foreign investors and visitors.

This double-edged strategy of promoting the city as a 'global metropolis' with 'local', 'exotic' characteristics has informed KD efforts since the 1990s, with efforts to attract global attention going hand in hand with those to 'sanitise' the city.[18] Areas considered dangerous for middle class Istanbulites and international visitors thus increasingly became targets for 'new' KD schemes. With their proximity to the historical centre of the city, such areas have also grown more attractive as loci for potential 'urban rent'.

THE GOVERNMENT'S KD METHODS AND THE SULUKULE COMMUNITY'S RESPONSE

At the level of urban governance, more programmatic KD efforts since 2000 have surpassed the cosmetic and clumsy efforts of the past. As a new approach to 'historic preservation and restoration', KD is a key policy for selling the city in its new image, one which aims to create synergies between Istanbul's aforementioned 'global' and 'local' characteristics. In 2005, on the request of the Justice and Development Party (AKP) government, the parliament passed the Urban Transformation Law[19] which allowed significant shortcuts to get around bureaucratic regulations. This legislation enables local governments to force owners to evacuate and renovate their historic buildings, either on their own or with the help of the Public Housing Administration or the municipalities.

Similar plans were rapidly implemented in Sulukule. Guided by place promotion efforts tied to an agenda of 'conservative gentrification' (whereby pro-AKP owners and investors reportedly bought up properties), KD projects are turning Sulukule (and others districts around the historic peninsula and elsewhere) into 'attractive', 'sanitised' neighbourhoods (Ünlü 2005). For Sulukule and other historic neighbourhoods, the municipally-owned Residential Development Planning Co. announced the demolition of rundown buildings[20] and construction in the model of 'Ottoman-Turkish architecture' to turn them into parts of a 'museum-city that could attract

10 million tourists a year' (Çavdar 2007). Note that such statements are reminiscent of other urban governance schemes around the world where selected locations within cities are revalorised.

Based on a nostalgic and selective reconstruction of the 'Ottoman-Turkish' past, the KD project is in line with the strategy of promoting the city as a 'global metropolis' with 'exotic' characteristics. The municipality's plan envisaged demolition of large sections of Sulukule, especially its most run-down buildings—whereby most landlords would be moved to housing developments away from the city centre, while tenants would be forced to relocate on their own terms. The plan also included the restoration of the 'most historically valuable' structures, the building of an underground car park, a boutique hotel, and a 'trade and cultural centre' that reportedly also includes a 'conservatory' for 'talented Sulukuleans'.[21]

Following the launch of this plan, the municipality began demolishing sections of Sulukule in 2007 with few or no prior warnings to residents. Professional organisations, NGOs and community groups criticised the urgency in which the decisions were taken and implemented, as well as the lack of participation by residents who will be left out of the negotiations between the municipalities and absentee landlords (Hallaç 2007; Ciravoğlu and İslam 2006). It was also noted that the plans would further harm Sulukule's social economy and solidarity-based networks.

Swift action by the national and local governments provoked new types of community organising, activism and alliance-building, as well as intra-community divisions vis-à-vis urban governance, to which we now turn.

ALLIANCES BUILT AND MISSED, COMMUNITY SPLIT: LESSONS FROM SULUKULE

Sulukule's residents had different reactions to the announcement of forced displacement. Especially those with backgrounds in the music, entertainment and tourism sectors who had worked in and/or owned entertainment houses resisted demolition. Many within this group also unapologetically called themselves Roma or Gypsy. They claimed that the project was planned without their participation and would eventually destroy their community, and believed the most effective way to fight 'non-participatory, destructive and neo-liberal' urban policies was to establish a community organisation and to strive to form broader alliances.[22]

With the help of numerous NGOs, academics, activists, architects, urban planners and others, the Sulukule Roma Culture and Solidarity Association was established in 2006. In a parallel development and in solidarity with the Sulukule Association, anti-demolition activists, NGO professionals, urban planners, architects, scholars and others established the Sulukule Platform. The Association and the Platform have jointly organised numerous activities, drafted alternative development and community-involvement plans and helped to increase the visibility of the issue both nationally and internationally.

In the eyes of the Sulukule Association and Sulukule Platform, plans for the boutique hotel, trade and cultural centre and conservatory meant little for the community as most tenants would be displaced.[23] Were the municipality to espouse a more participatory and transparent approach, they argued, local experience and potential could easily be integrated within tourism and entertainment-led development. The Sulukule Association and the Sulukule Platform thus jointly initiated a campaign to stop speedy demolitions and to revive the entertainment houses. They proposed renovating buildings without evacuating tenants and landlords and to allow residents to open new, government-regulated establishments.

The coalition of community-based organisations and NGO and activist groups stressed the need to protect and further develop Sulukule's tradition of entertainment. According to the coalition, demolitions would not only hurt the neighbourhood, but Istanbul as a city. The overall strategy showed not only the importance of forming broader alliances in the face of non-participatory urban governance practices, but also the usefulness of creating alternatives building upon existing social capital. The coalition launched '40 Days 40 Nights in Sulukule', a series of cultural, musical and advocacy events designed to draw public attention to the neighbourhood's plight. 'Do not destroy Sulukule, do not let Istanbul's joy be disrupted' emerged as the slogan of numerous campaigns, helping to associate Sulukule with a lively and socially meaningful tradition of entertainment.[24] The slogan indeed helped to put Sulukule on the map within wider social circles and to draw famous public figures (such as the legendary Turkish pop music star Sezen Aksu) to the anti-demolition campaign.

Here links with the critical infrastructure were crucial, in the sense that Sulukule's allies included numerous well-known figures in the media and cultural sectors. Numerous 'lifestyle' and 'gusto' columnists, film and music producers, cultural tourism experts, and advertisement and culture industry professionals became actively involved in assigning a new 'respectability' to the Roma identity in general and the neighbourhood in particular. This was in line with selective 'ethnicisation' in Istanbul's entertainment and tourism sectors, whereby the previously disdained Roma identity was re-invented and commodified through the promotion of 'world fusion' as a musical and cultural form. Contributions from members of the critical infrastructure thus constructed a new 'gaze' towards the Roma and Sulukule which helped to reach out to the middle and upper class residents of Istanbul.[25]

Indeed, concerts in Istanbul's experimental new clubs such as *Babylon* in Beyoğlu (the entertainment heartland of the city) brought together renowned groups such as Amsterdam Klezmer Band and Brooklyn Funk Essentials with Turkish Roma musicians and bands such as Galata Gypsy Band and Laço Tayfa. Similarly, the Platform organised numerous concerts by the Roma Orchestra of Sulukule to generate public interest.

While such efforts were usually successful and lauded by members of the critical infrastructure, others criticised them for developing an Orientalism of a different sort towards the Roma and reducing them to mere 'musical objects'.

The approach indeed had numerous problems. While the campaign helped to link Sulukule to its long entertainment tradition and attracted larger groups into the anti-demolition ranks, it also isolated Sulukule from other poor neighbourhoods targeted by KD plans and made it more difficult to establish ties of solidarity with them. While the Sulukule Association takes part in the Istanbul Platform of Neighbourhood Associations—a large network of working class, 'ethnic' and poor neighbourhood organisations established to counter non-participatory urban governance schemes and policies—there is visible strain and discord between the Sulukule and other neighbourhood groups.

According to some community organisations in working class, right-leaning or socially conservative neighbourhoods, Sulukule receives privileged treatment in the media and public debate (partly due to the contribution of members of the 'critical infrastructure' as discussed above), making it appear that KD plans only hit Sulukule. This reportedly damages or distorts efforts to inform the public about the true scale and potential dangers of neoliberal urban governance policies. They also argue that the emphasis on Sulukule's entertainment tradition and/or the Roma identity is perceived negatively by socially conservative residents who may otherwise join the ranks of anti-demolition organisations in numerous other neighbourhoods.

This indeed seems to be the case, especially in Fatih district, of which Sulukule is a part. Since Fatih is a socially conservative district, the municipality has been able to capitalise on Sulukule's 'notoriety' to gather public support for its KD plans. A group of Sulukule inhabitants wishing to distance themselves from the labels of Sulukule and/or being Roma (either for 'ethnic' and/or religious 'sensitivity' reasons) supported the KD project and publicly declared that 'Sulukule [indeed] needs to be cleaned up'. The municipality capitalised on their presence to strengthen its case that there is Sulukule-based support for the otherwise criticised project. Most belonged to the same group of Sulukuleans who had a negative view of entertainment houses and refused to call themselves Roma or Gypsy, and who instead emphasised their 'Muslim' and 'Turkish' roots.

The neighbourhood's conservative and Islamic-oriented residents indeed had strong ties with the local authorities in Fatih. Together they established the pro-KD Haticesultan-Neslişahsultan Association in Sulukule in 2006, which was instrumental in supporting the municipality's KD plans in public and decision-making circles. Their negation of the epithet Sulukule and everything it historically and socially stood for needs to be underlined here. This is instructive as community organising to counter non-participatory urban governance schemes provoked splits within the community and helped gather support for the same policies.[26]

CONCLUSION

Sulukule is an important and interesting case that demonstrates the complex interrelationships between the urban political economy, urban governance and urban culture. The case also illustrates the dynamic interplay between producers, consumers, members of the critical infrastructure and government, which explains the social relevance and spatial locus of an ethnically identifiable neighbourhood within a service and consumption-oriented urban economy. In this sense—and akin to the 'mixed embeddedness' approach (Kloosterman *et al.* 1999; Kloosterman and Rath 2001)—the case of Sulukule illustrates the need for situating entrepreneurship (here 'ethnic' rather than 'immigrant') within wider political, social and economic structures by taking into account the roles of government action and regulation, market dynamics and the cultural features of ethnic groups.

Important temporal and theoretical differences with other cases should also be noted here: while in the advanced economies of the West, and especially in continental Europe, the emergence of ethnic or immigrant neighbourhoods as sites of leisure (Rath 2005) is a recent phenomenon, Sulukule has a long history of being at the centre of Istanbul's leisure and entertainment economy. The Roma's migration history, shaped by imperial and republican demographic policies, facilitated their physical concentration in Sulukule, while the community's marginal, informal character contributed to the development of functional community networks and insulated the proclivity towards service and entertainment, which in turn led to the growth of an urban enclave identified with such professions.

Sulukule has historically been a locus of urban leisure consumption, a 'haven' in the face of conservative social pressures and a 'niche' alternative to top-down Republican-Kemalist cultural policies. Social pressures and authoritarian cultural policies helped to create an 'opportunity structure', a market for services that Sulukule entrepreneurs were able to capture. With their historical, social, economic, administrative and cultural marginalisation and supposed 'proclivity' for entertainment as well as their services-based community networks, only the Roma could supply these otherwise disdained services.

This certainly differs from other cases in post-industrial cities where the emerging service, leisure and tourism sectors made cultural diversity itself a commodity (Ley and Olds 1988; Kearns and Philo 1993; Lash and Urry 1994), providing opportunities for immigrant communities and entrepreneurs as well as for multicultural neighbourhoods (Rath 2007). The Sulukule community and its ethnic entrepreneurs did not capitalise on their human and social capital within such a permissive environment, but rather carved out a niche market where cultural difference was not encouraged. As government action and repressive social and cultural structures also discouraged cultural difference, the Roma's 'otherness' helped them create 'escapist' spaces such as entertainment houses which constituted the basis of their businesses.

Interestingly enough, just when Istanbul and Turkey's globalisation—and the rising (spending) power of the 'critical infrastructure' and consumers desiring 'exotic' experiences, goods and services—turned cultural difference and diversity into valuable commodities, moralising and criminalising governmental action and neoliberal urban transformation schemes destroyed the socio-economic basis of Sulukule's Roma community. While the local government had a clear agenda to sell a city boasting both 'global' and 'local' 'exotic' characteristics, a community that was ready and able to mobilise its human and social capital to revitalise its depressed urban neighbourhood was left out of the process, providing further evidence of the importance of government regulation for the viability of entrepreneurial initiatives even when other 'ingredients' of the equation are present. The case of Sulukule thus confirms that the reshaping of cities as places of consumption contains both opportunities (Fainstein and Judd 1999; Kearns and Philo 1993) and challenges for local economies and communities.

While government regulation and neoliberal 'place marketing without the people' were the main factors behind the Roma of Sulukule's eventual demise, inter-community divisions and alliances with members of the critical infrastructure, middle and upper class residents, national and international NGOs and international bodies (such as the UN, European Parliament, the European Commission and others) also posed challenges. The forging of alliances solely with 'outside' supporters created discord *within* the Sulukule community and *between* Sulukule and its 'natural' allies including other neighbourhood-based associations fighting against top-down and non-participatory urban governance schemes and practices.

The above points notwithstanding, Sulukule has shown that imaginative community responses based on local entrepreneurship and innovation, catering to the demand for 'cultural diversity sought by a local yet cosmopolitan clientele' (Zukin and Kosta 2004) and international tourists, and reviving local communities and economies by forging transparent and participatory mechanisms of urban governance remain as possibilities. While by late 2008 KD plans had demolished much of Sulukule and such chances had become increasingly ephemeral, Sulukule and its entertainment houses continue to narrate a hopeful and heartening story.

ACKNOWLEDGMENTS

The authors would like to thank Jan Rath, Ayşe Çavdar, Baki Tezcan, Savaş Arslan, Zeynep Gönen, Filiz Babalık and Ogün Duman for their important comments and contributions to this chapter. The responsibility for all possible omissions and mistakes are obviously ours.

NOTES

1. *Kentsel Dönüşüm* (KD), as referred to in governance and NGO circles, translates as 'urban transformation'. In its formulation and implementation, it more closely resembles urban regeneration/renewal, with an emphasis on (selective) historical preservation and restoration.
2. Marsh (2006) dates this history back to the twelfth century, while Kalafat and Aydın (2007) trace it to the eleventh century.
3. While in modern Turkish usage *millet* is a 'nation', in the Ottoman Empire it meant 'confessional community', loosely based on *shari'ah* law's acknowledgement of the rights of non-Muslim groups. In the Ottoman Empire, at first only the largest and most visible non-Muslim groups such as the Orthodox Greeks, Gregorian Armenians and Jews were recognised as *millet*s and accorded a degree of autonomy. For discussion of the *millet* system, see Braude and Lewis (1982). In the empire, notions around 'ethnicity' only appeared in the late nineteenth century, under the influence of westernisation, modernisation and the advent of French Revolution-inspired nationalisms. In the Republic of Turkey, the notion of ethnicity remained highly controversial, mainly due to the insistence on political, administrative, economic and cultural centralism and fear of ethnic separatism. For these reasons, numerous non-Turkish groups, including the Roma, carefully avoided identifying themselves as an 'ethnic' or 'minority' group.
4. Some historical accounts claim that the Ottoman Army's celebrated marching band was first established by Istanbul's Roma. Ottoman cabinet minister Ali Rıza Bey (1842–1928), who wrote in rich and colourful detail on Istanbul's quotidian life, gives numerous examples of the Roma's domination of entertainment. See Balıkhane Nazırı Ali Rıza Bey (2001) *Eski Zamanlarda İstanbul Hayatı* (*Life in Istanbul in Old Times*) assembled and edited by A.Ş. Çoruk from his writings in the late 1910s through to the mid-1920s.
5. *Zurna* (a clarion of sorts) virtuoso Şahin was among the official music instructors of the palace. Other famous *zurna* virtuosos, such as Emin Tanınmış and İbrahim Özüfler, were recruited as gramophone artists and their recordings became available to a larger public. *Tambur* (an Ottoman lute) virtuoso Cemil was reportedly taught by Arap Emir, a famous Roma musician from Sulukule (see Kalafat and Aydın 2007).
6. Works by Evliya Çelebi, Ali Rıza Bey and other literary figures and by satire writers such as Ahmet Rasim (1922 [1991]) richly describe the role of the Roma in Istanbul's entertainment sector and in daily life.
7. Interestingly enough, during the republican period the Roma were referred to in legal texts, administrative jargon and popular parlance as 'swarthy citizens' ('*esmer vatandaş*').
8. Entertainment houses were also referred to in Turkish as *Devriye Evleri*, a loose translation for which is 'Houses of Rounds' or 'Houses of Turns', referring to the 'rounds' of entertainment taking place there or musicians/dancers 'taking turns' in their performative acts.
9. See, for example, Aytar and Keskin (2003). Tekelioğlu (1996) argues that music was among the key instruments in the modern Turkish state's attempt to fashion a new sort of citizen and nation-state. He cites the banning of non-Western, popular and Ottoman-originated musical forms from state radio, suppression of the public display of religious/confessional music and dance, promotion of polyphony, 'Turkification', standardisation, archivisation and orchestrialisation of various types of Anatolian folk music and the launching of state-sponsored ballroom dances.

10. *Oturak âlemi* literally means 'stool session', referring to the stools that were sat on. *Oturak âlemi* was indeed reminiscent of the circular performances at the *Eğlence/Devriye Evleri*. In mainly Kurdish south-eastern Anatolia, comparable sessions were organised under the rubric of *sıra gecesi* ('Night of Turns', again denoting circular notions of musicians/guests taking turns). However, alcohol and female dancers and escorts were not allowed.
11. Here Michel de Certeau's (1980) analyses of 'strategies' and 'tactics' in quotidian life may be instructive. For a study of the social 'protest' functions of popular and folk music in the Ottoman Empire and the Republic of Turkey, see Gündoğar (2005).
12. Numerous films have strengthened the moralistic and criminalising popular gaze towards the Roma and Sulukule—the 'exotic and/or ridiculed/criminalised Other'. In the early 1980s, the *Gırgıriye* (a play on words based on references to jocularity and noisy entertainment) series of films on two fictionalised rival Sulukule entertainment families shaped the neighbourhood's popular perception as a dirty, seedy, urban ghetto populated by ignorant, noisy, lazy, lying, constantly fighting yet crudely 'fun' residents. Even today, entrepreneurs and musicians in Sulukule remain highly critical of *Gırgıriye*-type movies and TV series and underline the fact that they all give a distorted, negative image of their neighbourhood.
13. Besides appearing in countless national movies, Sulukule was one of the locations for the shooting of the James Bond movie *From Russia with Love* (1963), increasing the neighbourhood's attraction for tourists.
14. Roma musicians such as Hüsnü Şenlendirici, Adnan Şenses, Kibariye and Sibel Can are reportedly among them.
15. Twenty-nine buildings were reportedly demolished in Sulukule, though it is unclear if any were entertainment houses.
16. Note that Sülayman Ulusoy was known as *Hortum Süleyman* (Süleyman the Hose) as he reportedly beat those in custody with a hose (Soykan 2000). Ulusoy was later instrumental in 'cleaning up' Cihangir neighbourhood with its transvestite residents, before asking for early retirement when his practices were reported in the media.
17. Istanbul was transformed from a base for large-scale manufacturing into Turkey's globalising centre for banking and finance (see for example Keyder and Öncü 1994). The city has also increased its hold over the Turkish economy and has become the preferred site for multinational corporations entering the Turkish market. Tourism has also played a large part in globalising Istanbul's reach. See for example Pérouse 2000.
18. This first crystallised with the city's hosting of the UN Conference on Human Settlements (Habitat II) in June 1996. Prior to this, street children, beggars, Roma flower sellers, transvestites, sex workers and others were transferred outside the central areas, avowedly to minimise contact with international participants. For discussion see Selek (2001). Similar 'sanitisation' policies continued, preceding the NATO conference in 2004.
19. Law No. 5366, sometimes dubbed the 'Beyoğlu Law' or the 'Historic Peninsula Law', mainly targets those Istanbul neighbourhoods close to the city centre which became 'fashionable' over the last decade or so.
20. In July 2006 the government issued a decision to 'urgently expropriate' parts of Sulukule and other neighbourhoods. In Sulukule this entailed the demolition of about 571 houses, the majority inhabited by Roma. Hundreds of families will be displaced; approximately half who are tenants will not receive replacement housing. Critics argue that this top-down, 'cosmetic' policy will remove the urban poor to pave the way for tourism-, big business- and economic rent-driven processes of inner-city gentrification (see Hallaç 2007).

21. For rather celebratory accounts of the plan, see pro-AKP dailies *Yeni Şafak* and *Zaman*'s various news stories and op-ed columns in 2007 and 2008. See for example 'Sulukule'de Yeni bir Hayat Başlıyor' (A New Life Starts in Sulukule) at http://yenisafak.com.tr/Cumartesi/?t=10.11.2008&i=149027. For critical approaches see Oral (2007); Kırca-Schroeder and Somersan (2008).
22. A group of Roma gathered around the well-known community leader Şükrü Pündük, a musician who previously owned entertainment houses. Pündük was then running a tea house in the neighbourhood, which became a focal point for organising. While confident of the strength of his own community, Pündük was concerned about forging alliances with outside groups to continue fighting demolitions. In a focus group study he stated: 'The state officials are not taking us seriously and they create obstacles all the time. We need non-Roma people or organisations to back us up'.
23. According to the OECD (2008), the Fatih and Beyoğlu districts lack comprehensive regulatory and financial frameworks that integrate urban regeneration, housing and urban development objectives and the protection of low-income residents. The review addresses the current situation in Sulukule and criticises local authorities for not considering the consequences of breaking up community networks and not compensating vulnerable residents. According to the review, within the limits of Neslişah and Hatice Sultan, 571 owners and 391 tenants were invited to participate in the program aimed at protecting historical buildings and improving urban design. Under the current negotiations, tenants without property are simply invited to leave the neighbourhood, while 571 owners receive a lump-sum compensation of YTL 500/m^2 (around US$ 350/m^2). Fatih municipality's negotiations framework has since late 2005 excluded the majority of residents.
24. In 2007 and 2008, the Sulukule Platform organised a workshop bringing together Mimar Sinan Fine Arts University and London School of Economics scholars and students to delineate an alternative local development plan. '40 Days 40 Nights Sulukule Events' included concerts by Roma musicians, painting exhibitions, panel discussions, film screenings and conferences on the streets in Sulukule and at various universities, while international celebrities such as musician Manu Chao, members of the music band Gogol Bordello and Roma film director Tony Gatlif visited the neighbourhood. Through the work of the Platform, members of the United States House of Representatives, the European Parliament, the UN Special Rapporteur on the right to adequate housing as a component of the right to an adequate standard of living, and the UN Independent Expert on Minority Issues sent letters warning the government to espouse a more participatory approach in Sulukule. The Sulukule Association meanwhile initiated a petition campaign and submitted the petitions to Prof. Zafer Üskül, an AKP MP and the Chair of the Human Rights Commission at the Turkish Grand National Assembly.
25. For a discussion of selective 'ethnicisation' and the re-invention of the Roma, see Aytar 2007.
26. In the process of negotiations with the municipality, some residents changed sides. Those who thought they would benefit from the demolition economically and/or socially worked closely with the municipality and began selling their properties.

REFERENCES

Akçura, G. (2007) 'Sulukule kültürünün tarihi ve geleceği' <The history and future of Sulukule culture>, *Yeni Mimar*, April: 66–68.

Aytar, V. (2007) 'Caterers of the consumed metropolis: Ethnicized tourism and entertainment labourscapes in Istanbul', in Rath, J. (ed.) *Tourism, Ethnic Diversity and the City*. New York and London: Routledge, 89–106.

Aytar, V. and Keskin, A. (2003) 'Scuffling and intermingling sounds in a fragmented metropolis: constructions of spaces of music in Istanbul', *Géocarrefour: Revue de Géographie de Lyon*, 78(2): 147–57.

Beler, T.A. (1946) *Beyoğlu Piliçleri <Chicks of Beyoğlu>*. Istanbul: no publisher given.

Braude, B. and Lewis, B. (eds) (1982) *Christians and Jews in the Ottoman Empire: The Functioning of a Plural Society*. New York and London: Holmes & Meier Publishers.

Çavdar, A. (2007) 'Habersiz, derinden ve usul usul: Tarlabaşı'nı sürmek' <Deeply, quietly and stealthily: Forcing Tarlabaşı out>, *Istanbul Kent Kültürü Dergisi <Istanbul Journal of Urban Culture>*, 58: 48–51

Ciravoğlu, A. and İslam, T. (2006) 'Beyoğlu ve Soylulaştırma' <Beyoğlu and gentrification>, *Mimarist*, 21(Fall): 6–9.

de Certeau, M. (1980) *L'invention quotidien: Arts de faire*. Paris: Folio.

Fainstein, S. and Judd, D. (1999) 'Global forces, local strategies and urban tourism', in Fainstein, S. and Judd, D. (eds) *The Tourist City*. New Haven and London: Yale University Press, 1–17.

Foggo, H. (2007) 'Sulukule Sulukule olalı böyle zulüm görmedi' <Sulukule has never, ever witnessed such cruelty>, *Bianet*, 12 January. <http://bianet.org/bianet/bianet/90131-sulukule-sulukule-olali-boyle-zulum-gormedi> (accessed 15 June 2009).

Gündoğar, S. (2005) *Muhalif Müzik: Halk Şiirindeki Protesto Geleneğinden Günümüz Politik Şarkılarına <Music of the Opposition: From Protest Tradition of Folk Poems to Today's Political Songs>*. Istanbul: Derin.

Hallaç, M.A. (2007) 'Kentsel dönüşüm projeleri ayrımcıdır' <Urban transformation projects are discriminatory>, *Bianet*, 31 March. <http://bianet.org/bianet/bianet/93927-kentsel-donusum-projeleri-ayrimcidir> (accessed 15 June 2009).

Halter, M. (2000) *Shopping for Identity: The Marketing of Ethnicity*. New York: Schocken Books.

İnalcık, H. (ed.) (1997) *Economic and Social History of the Ottoman Empire, 1300–1914*. Cambridge: Cambridge University Press.

Kalafat, G. and Aydın, F. (2007) 'Sulukule'. <http://istanbulmap.org/articles/Sulukule.doc> (accessed 15 June 2009).

Kayaoğlu, G. (1996) *Evliya Çelebi Seyahatnamesi <Book of Travels of Evliya Çelebi>*. Istanbul: Yapi Kredi Yayinlari.

Kearns, G. and Philo, C. (eds) (1993) *Selling Places: The City as Cultural Capital Past and Present*. Oxford: Pergamon Press.

Keyder, Ç. and Öncü, A. (1994) 'Globalization of a Third-world metropolis: Istanbul in the 1980s', *Review*, 17(3): 383–421.

Kırca-Schroeder, S. and Somersan, S. (2008) 'Eğlence Evlerinden Yıkıma: Sulukule Mikrokozmosu' <From entertainment houses to demolition: A Sulukule microcosm>, *Istanbul Kent Kültürü Dergisi <Istanbul Journal of Urban Culture>*, 62: 88–91.

———. (2007) 'Resisting eviction: Sulukule Roma in search of right to space and place', *The Anthropology of East Europe Review: Central Europe, Eastern Europe and Eurasia*, 25(2): 96–107.

Kloosterman, R. and Rath, J. (2001) 'Immigrant entrepreneurs in advanced economies: Mixed embeddedness further explored', *Journal of Ethnic and Migration Issues*, Kloosterman, R. and Rath, J. (eds) *Special Issue on 'Immigrant Entrepreneurship*, 27(2): 189–201.

Kloosterman, R., van der Leun, J. and Rath, J. (1999) 'Mixed embeddedness: (In) formal economic activities and immigrant business in the Netherlands', *International Journal of Urban and Regional Research*, 23(2): 253–67.

Lash, S. and Urry, J. (1994) *Economies of Signs and Space*. London: Sage.

Ley, D. and Olds, K. (1988) 'Landscape as spectacle: World fairs and the culture of heroic consumption', *Environment and Planning D: Society and Space*, 6: 191–212.

Marsh, A. (2006) *The Gypsy Diaspora: Ethnicity, Migration and Identity*. Istanbul: Consulate General of Sweden in Istanbul and the International Romani Studies Network.

OECD (2008) *OECD Territorial Reviews: Istanbul, Turkey*. Istanbul: OECD.

Oral, F. (2007) '40 kere söyleyeceğiz: Sulukule yaşayacak!' <We'll say this, forty times over: Sulukule will survive!>, *Istanbul Kent Kültürü Dergisi <Istanbul Journal of Urban Culture>*, 60: 124–26.

Pérouse, J.-F. (2000) 'L'internationalisation de la métropole stambouliote', *Varia Turcica*, 34: 153–79.

Rath, J. (ed.) (2007) *Tourism, Ethnic Diversity and the City*. London and New York: Routledge.

———. (2005) 'Feeding the festive city: immigrant entrepreneurs and tourist industry', in Guild, E. and van Selm, J. (eds) *International Migration and Security: Opportunities and Challenges*. London and New York: Routledge, 238–53.

Rasim, A. (1922 [1991]) *Dünkü İstanbul'da Hovardalık: Fuhş-i Atik <Womanising in Yesterday's Istanbul: Old Harlotry>*. Istanbul: Arba.

Rıza Bey, A. (2001) Çoruk, A.R. (ed.) *Eski Zamanlarda İstanbul Hayatı*. Istanbul: Kitabevi.

Sabanci, G. (2008) 'Suluuyle'de Yeni bir hayat başliyor.' <http://yenisafak.com.tr/cumartesi/?t=10.11.2008&i=149027> (accessed 10 August 2008)

Selek, P. (2001) *Maskeler Süvariler Gacılar: Ülker Sokak, Bir Kültürün Dışlanma Mekânı <Masks, Cavaliers and Broads: Ülker Street, a Space of Marginalisation of a [Sub]culture>*. Istanbul: Aykırı.

Soykan, T. (2000) 'Sulukule'de "Hortum" Bayramı' <'Hose' Feast in Sulukule>, *Radikal*, 3(June): 6.

Tekelioğlu, O. (1996) 'The rise of a spontaneous synthesis: The historical background of Turkish popular music', *Middle Eastern Studies*, 32(2): 194–218.

Ünlü, A. (2005) 'İstanbul'un görünmez merkezi Tarlabaşı'nın görünenleri' <The unseen centre of Istanbul: Tarlabaşı and what is now seen there>, *Mimarist*, September: 48–52.

Yeni Şafak (2008) 'Sulukule'de Yeni bir Hayat Başlıyor' <A new life starts in Sulukule>. <http://yenisafak.com.tr/Cumartesi/?t=10.11.2008&i=149027> (accessed 15 June 2009).

Yılgür, E. (2007) 'İstanbul çingeneleri: Karşılıksız bir aşk hikayesi' <Gypsies of Istanbul: a non-reciprocal love story>, *Istanbul Kent Kültürü Dergisi <Istanbul Journal of Urban Culture>*, 57: 34–39.

Young, T. (1963) *From Russia with Love*. United Artists.

Zukin, S. and Kosta, K. (2004) 'Bourdieu off-Broadway: Managing distinction on a shopping block in the East Village', *City & Community*, 3(2): 101–14.

6 When Diversity Meets Heritage

Defining the Urban Image of a Lisbon Precinct

Catarina Reis de Oliveira[1]

INTRODUCTION

While some cities recognise and even promote immigrant neighbourhoods in their overall branding, in other cities such strategies enter into conflict with other urban identities. This contribution seeks to understand why a particular Lisbon neighbourhood, Mouraria, has *not* developed into an ethnic tourist attraction. The case of Mouraria is particularly intriguing as recent immigrants have settled in an area with a rich material and symbolic heritage drawn from different cultures, languages, populations and religions.

This chapter examines the contradictions inherent in spatial narratives and between policies promoting multiculturalism and place heritage. As we will see, it can be more difficult to promote cultural diversity as the centrepiece of a district's image when it is located in the historic centre of the city, or when the district itself represents the city's identity. The role Mouraria plays in Lisbon's identity has become a fundamental barrier to those who support the idea of urban cultural diversity. The eight centuries of history experienced in Mouraria reveal the contrast between the medieval imaginary—rooted in the genesis of the city and the Christian origins of Portugal—and today's multicultural reality.

We begin with a review of the literature on the commodification of ethnically diverse districts, highlighting in particular the role of political structures and regulation in this process. The first section also contains a brief description of Mouraria, its location in the city of Lisbon, its residents and local entrepreneurs. The subsequent sections then examine how different—and conflicting—politics of culture and ancient heritage can be advocated by different actors targeting the same neighbourhood. Three main discourses exist in Mouraria: the first relates to the past, a nationalism of ancient heritage. The second concerns the present diversity of the neighbourhood; the third emphasises the dynamism of the local economy. Inherent in these discourses is the question of entitlement: who can legitimately define Mouraria's place image?

THE COMMODIFICATION OF ETHNICALLY DIVERSE DISTRICTS

Immigrant districts in several countries have become symbols of cultural diversity and objects of local pride, place marketing and tourism (Halter 2000). In certain cases, ethnically diverse districts have even come to represent the identity of the cities themselves (Hall and Rath 2007). Through pronounced ethnic symbolism, local authorities have 'artificially' kept alive some of these areas as urban attractions—even after immigrants had left the area (Collins 2007).

As researchers have shown in other contexts, the 'opportunity structure' plays an important role in the commodification of ethnically diverse districts (Zukin 1991; Collins 2007; Rath 2007a). The mere presence of immigrants and ethnic entrepreneurs is insufficient for neighbourhoods to emerge as tourist attractions or to be recognised by outsiders as ethnic precincts. As Rath (2007a: 10) argues, the transformation of minority neighbourhoods into places of leisure and cultural consumption is no straightforward process, and depends on the interplay of several factors: the social infrastructure (including immigrant entrepreneurs and immigrant residents), the proliferation of small businesses offering ethnic goods and services, the critical infrastructure (including consumers, cultural producers who influence public taste, business associations, tourist boards and the local government) and the regulatory framework (including the area's accessibility, the prevalence of crime, regulations relating to shop fronts and so on).

It is significant that this transformation did not fully take place in Mouraria. We thus need to examine other factors, namely the interference of Mouraria's heritage and its material attributes in the place *imaging*[2] of the neighbourhood. In analysing the creation of place images that feature cultural diversity, the Mouraria case highlights that—together with the material processes that produce the urban landscape—there can be centuries of heritage that impose specific constraints, ultimately leading to diverse spatial narratives.

As demonstrated by Hall and Rath (2007: 19), the promotion of place is contingent on what members of the 'critical infrastructure' consider marketable attractions. Certain ethnic concentrations may be attractive to overseas markets but not to local ones, and vice versa. Although cultural diversity attracts domestic consumers to Mouraria, the tourist office and local authorities have chosen a different narrative to sell to international visitors: Mouraria as the setting of Lisbon's birth. Because of its location and role in Lisbon's history, Mouraria is a fascinating case to analyse the interaction of attractive 'otherness' with centuries of heritage landscape.

Mouraria's role in Lisbon's history—and in the history of Portugal—became the cornerstone of the debate on the transformation of an ethnic district into a place of leisure and cultural consumption. Zukin (1995: 7) argues that

> building a city depends on how people combine the traditional economic factors of land, labor, and capital. But it also depends on how they manipulate symbolic languages of exclusion and entitlement. The look and feel of cities reflect decisions about what—and who—should be visible and what should not, on concepts of order and disorder, and on uses of aesthetic power.

Thus it is important to identify the arguments that, independently of their authenticity in claiming culture or identity, have the strength to influence public debate, neutralise counter-arguments and mobilise supporters (also argued by Gotham 2007: 12–13).

Although Mouraria is an ethnically diverse neighbourhood with a high concentration of immigrant entrepreneurs, it did not become a multicultural emblem for the city of Lisbon. Symbolic branding instead focused on Mouraria's medieval character, leading to attempts to conceal how the area is actually used at present. The primary focus here is on understanding the problematic process of Mouraria's ethnic commodification as well as the diverse urban imaging strategies that informed its non-consensual selection as a symbol of Lisbon's cultural diversity.

The Mouraria case further demonstrates the influence policy can have on the consolidation or absence of a neighbourhood's ethnic identity or identifiable 'otherness'. Research has shown how local authorities can maximise the advantages of the presence of visible immigrant communities, thereby demonstrating the economic value of cultural diversity for urban development (Shaw *et al.* 2004; Rath 2007b). This chapter, however, seeks to show how political influence can be defining in the opposite direction, constraining those economic and social interests that favour the neighbourhood's transformation into a place of leisure, diversity and cultural consumption.

Harvey (2003) has illustrated that identity politics may be reflected in contemporary descriptions of ancient sites. The arrival of immigrants to this old Lisbon neighbourhood and the pressure exerted by the new ethnic infrastructure to transform the area into an ethnically labelled attraction disrupted older spatial narratives, city myths and national identity. Tilley (2006: 19) has further examined heritage landscapes as memories of a nation's past—necessary to preserve identity as a counterpoint to modernity. In an attempt to freeze time, Mouraria's 'city myths' were recycled and 'artificial' symbols were introduced in order to provide a traditionally 'authentic' experience of the medieval. These contrast with the lively immigrant restaurants, kiosks and shops (advertised in foreign languages) and the immigrant users of public space. Thus faced with the preservation of eight centuries of heritage in Mouraria, immigrants had an uphill struggle to transform the neighbourhood into a multicultural and cosmopolitan district.

It could be argued that in cities with a long tradition of immigration, ethnic districts are often part of the nation's heritage, while in cities in new immigration and Old World countries, commodification for tourism implies competition between more recent settlements and other attractions linked to centuries of heritage. The commodification of ethnically diverse districts thus takes place within broader contexts; Yeoh (2005: 946) suggests that the colonial context from which some cities recently emerged necessitates their cultivation of national identity and pride.

Urban cultures and their symbolic economies are not limited to material practices, but co-exist with subtle strategies of social differentiation at the micro level (Zukin 1995: 11). Although the Mouraria case contains both economic actors (immigrant entrepreneurs) and political actors with the power to shape public culture through stone and concrete (the city government), it shows that public space is socially constructed and that people themselves enjoy agency in defining the city's image.

MOURARIA: THE CONSOLIDATION OF AN ETHNIC PRECINCT?

Mouraria is one of Lisbon's oldest neighbourhoods, located in the historic inner city (see Figure 6.1). The neighbourhood's age entails numerous disadvantages: poor housing and living conditions, and a high incidence of poverty and unemployment.

During the twentieth century, Mouraria became a run-down district with an aging population. Significant parts of the neighbourhood were condemned and razed to the ground, bequeathing to the area an air of destruction, abandonment and decay. The damage done to the local economy reinforced the neighbourhood's social, functional and physical marginalisation. As a result, many inhabitants and shopkeepers moved to other residential and commercial areas in Lisbon.[3]

In two surveys conducted in the late 1990s to assess the attitudes of Lisboetas towards 26 areas of the city, Mouraria was rated the seventh most dangerous to walk in during the day (Esteves 1998). Although it was considered average in terms of danger, 73 per cent of respondents believed Mouraria lacked effective policing (Almeida 1998). Examining actual crime statistics, however, produces a different geography of security as Mouraria does not suffer from a particularly high crime rate. Esteves suggests that feelings of insecurity are fuelled by other factors including poor housing, the socio-economic characteristics of its residents, and the prevalence of prostitution and drug trafficking in surrounding areas.

It was precisely in this context of a run-down and devalued urban area that immigrants began to settle, revitalising the neighbourhood's public spaces and abandoned commercial areas. From 1981 to 2001, the number of residents in Mouraria declined by 54.8 per cent. This decline would have been even greater had immigrants not moved in. Countering the trend among Portuguese residents, the number of immigrants increased by 66.8 per cent between 1981 and 2001.

Figure 6.1 Location of Mouraria in Lisbon. *Map designed by Ana Sala.*

The first immigrant groups to invest in the neighbourhood, sharing the space with local Portuguese entrepreneurs, were ethnic Indians from Mozambique—both Hindus and Muslims. In smaller numbers, Africans also began investing in the area, setting up hair salons, cosmetics shops, crafts shops and ethnic music stores. Chinese entrepreneurs arrived in Mouraria almost 15 years later, following two extraordinary regularisation processes for undocumented immigrants in Portugal in 1992–93 and 1996.[4] Their strategy was to occupy the area's shopping centres and to invest in clothes shops selling Asian goods, ethnic supermarkets, gift shops and wholesale businesses. Over the years, and due to intense competition, Chinese entrepreneurs became particularly adept at lobbying, increasing both supply and demand for their goods in the neighbourhood.

It is important to acknowledge that immigrant entrepreneurs did not create the local market; Mouraria has been a wholesale district for grocery shops, clothes shops and restaurants for decades. However, immigrants arrived to revive and modernise the local market just when some of these businesses were entering decline, leaving more and more shops vacant.[5] Over time, the local market experienced an outflow of Portuguese and certain immigrant entrepreneurs, forming what Kloosterman and Rath (2001: 195) have dubbed *vacancy-chain* opportunities for newcomers. Mouraria is a typical vacancy-chain market. As Kloosterman and Rath point out, this kind of market is characterised by the succession of different immigrant groups, low barriers to entry and narrow profit margins due to intense competition with low-value-added products that soon saturate the market. Mouraria's immigrant entrepreneurs generally do not target consumers demanding ethnically diverse experiences. Their main clients are retail businesspeople of Roma, Indian, Chinese and Portuguese origin and consumers looking for inexpensive products.

The arrival of immigrants transformed Mouraria's social environment, from one containing ageing Lisboetas to one containing people of different ages and backgrounds. Martim Moniz Square, located at the centre of Mouraria's two main shopping centres and commercial streets, became a meeting point for immigrants, often just to hang out near the small outdoor cafés. During the day, immigrant children play in the square. The opening hours of immigrant shops also began determining the daily movement of people: during the day, the area is full of shoppers, residents, entrepreneurs, small business suppliers, street traders and migrant workers and their children. At night the area assumes a different character, frequented by prostitutes, drug addicts and homeless people.

While it is clear that immigrants over the past two decades have become a significant presence in Mouraria and have revitalised the local market, the area has not become a site for leisure and ethnic tourism. The following section ventures to explain why this is so.

THE POLITICS OF CULTURE AND THE POLITICS OF ANCIENT HERITAGE

Numerous actors and stakeholders have contributed to three co-existing and sometimes conflicting discourses on Mouraria. One discourse focuses on city heritage, the second on cosmopolitan diversity, and the third on entrepreneurship and economic dynamism. Analysis of these three discourses sheds light on recent politics in the neighbourhood and its attendant tensions.

The Heritage Discourse

Mouraria's heritage dates back to the medieval period. Around the year 711, Lisbon was taken over by Moors from North Africa and the Middle East. Under Moorish rule, the city flourished with a diverse population of Christians, Berbers, Arabs and Jews, with Arabic as the official language and Islam the official religion. Many houses and streets of the period survive today while Arabic place names also remain. It was not until 1147 that the city was 'reconquered' by Christians, commanded by the first king of Portugal, Afonso I, with the assistance of the Crusaders. Mosques were converted into churches and the population was gradually either converted to Christianity or banished. King Afonso I decreed that Moors must live outside the castle walls, defining Mouraria's identity as a segregated area for the defeated Moors. This history can still be seen today in the labyrinthine pattern of the streets and in the dense urban constructions, though many places have been renamed after new heroes.

Several regeneration plans have been implemented in Mouraria since the early 1930s. That of the 1980s is of particular significance as it was being implemented just as the first immigrants were arriving in the neighbourhood. It foresaw the rehabilitation of an 115,354m^2 area, with the creation of new commercial spaces, offices, houses and parking lots.

In 1985 the Lisbon city government opened a local office in Mouraria to promote the district's regeneration. The project entailed re-interpretation of what was considered historical and traditional: Lisbon's history was acknowledged only after the conquest of the Moors, thus overlooking the prior Moorish influence. Over the years, the technical support office wielded significant influence over local life. Although its role principally concerned the area's urban regeneration, it also intervened in economic activities, social networks and in the dynamic definition of local identities, linking the district to the medieval period (Menezes 2004).

Several developments in the late 1990s reflected the priorities of the city government in defining the area's identity. The Socorro metro station was renamed Martim Moniz station,[6] while the square was renovated with fountains, small replicas of military towers and statues of Crusaders, which were also displayed on the metro station platform. Based on the story of the first king of Portugal conquering the city from the Moors, these additions projected a heroic image for the nascent nation, celebrating great deeds as icons of national (and city) heroism.

Nationalism is legitimised by subtle signs such as the reinforcement of place images representing the nation-building myth, or in this case the city-building myth (Harvey 2003: 463; Johnson 1999: 188). Of particular relevance is the city's material culture, the abundance of objects and images that make up its urban space. In the absence of more personal contact, it is in the landscape that visitors organise their perceptions of the city (Wells 2007: 137). In the Mouraria case, the restoration of historical architectural detail, the introduction of new symbols and the restoration of old myths and heroes all reflect the city government's assertion of its role in defining the neighbourhood's identity—although immigrants had already appropriated the area.

Zukin (1987: 135) argues that 'the ideology of historic preservation facilitates the removal of a pre-gentrification population', a population considered incongruous to the district's 'authenticity'. The introduction of the heritage discourse indeed followed the realisation that the district's past was not immediately visible and was being obscured by daily life. Following Zukin's argument on urban symbolic economies, emphasising the conquest of the Moors is a powerful means of controlling the city and defining who belongs in specific places (Zukin 1995: 1–2). With the signs of the city's *reconquista* clearly at loggerheads with the signs of immigrant presence, Mouraria became an arena of competition over identity and symbolic appropriation.

The official websites of Portugal and Lisbon Tourism do not refer to the contemporary multicultural character of Mouraria. They instead focus on the city-building myth and the celebrations of the Patron Saints, emphasising Mouraria's Christian identity. As Harvey (2003) argues, heritage is a social process and an instrument of cultural power.

The performance of specific memories can generate conflict. As Jarman argues, the changing needs and circumstances of the present can stimulate both the re-evaluation of the past and the creation of new memories that conflict with emergent understandings of peoples' daily lives, while the pressure of providing continuity in the face of change 'makes it almost impossible to wipe the slate clean and begin again' (Jarman 1997: 5). As Zukin argues, the 'urban imaginary' that defines a place image is influenced by numerous actors: city governments, entrepreneurs, consumers, the media and daily as well as once-off visitors.

The Diversity Discourse

Eight centuries after the conquest of Lisbon from the Moors, citizens with diverse religions, mother tongues and cultural practices have returned to Mouraria. The district once again boasts an inter-cultural texture: immigrant entrepreneurs have introduced new products to the local market and are concentrated in several ethnic shopping centres and streets, alongside the traditional local traders. With this influx have come new consumers and new users of public space, while mosques have returned to the neighbourhood.[7]

The media now acknowledges the Mouraria neighbourhood as Lisbon's 'diver-city' where visitors can buy low-cost products and experience new flavours and aromas.[8] Images of foreign cultures are visible in the shop windows where immigrant entrepreneurs advertise their products and demands for labour in Creole, Chinese and Hindi, representing the 'ethnic personalisation' of space (Harney 2006).

Particular cultural events can be important forms of place-making when they connect urban space with ethnic attachment (Harney 2006: 30). The now fashionable ideology of ethnic diversity has informed a number of initiatives in Mouraria. To combat the negative stereotyping

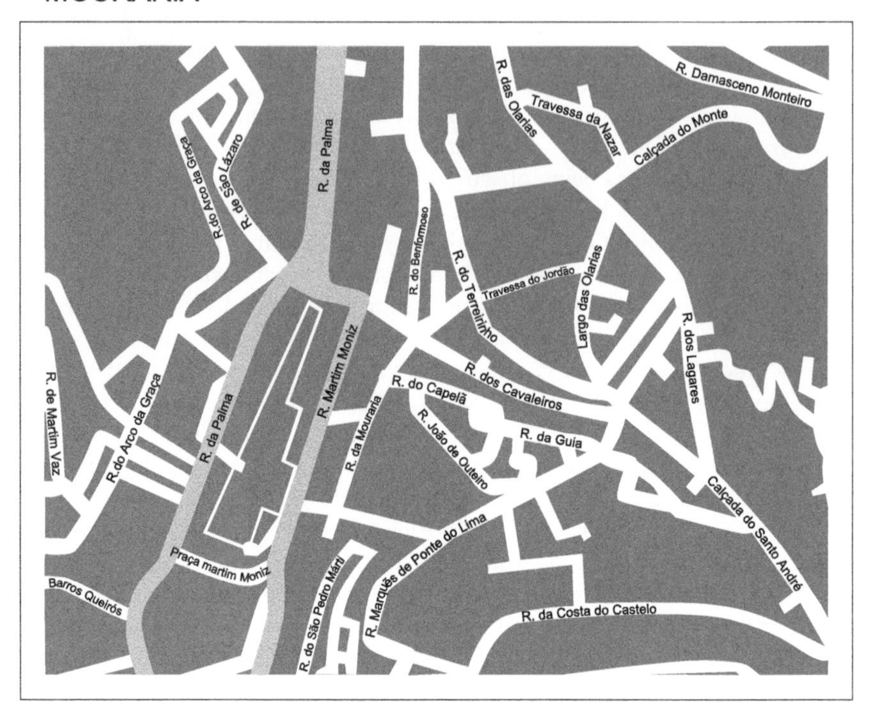

Figure 6.2 Map of Mouraria. *Map designed by Ana Sala.*

of the neighbourhood, the city government implemented its Multicultural and Cosmopolitan Policy.[9] The first multicultural carnival in Martim Moniz Square took place in 1998, with Brazilian, Chinese, Cape Verdean and Mozambican parades. Chinese New Year was celebrated in 1999.[10] Mouraria also hosted the *Festival of Diversity* in June 2003, and again in June 2006, attracting around 4,000 visitors to the square. In contrast to festivals that had taken place elsewhere in the city, these two occasions stimulated the direct involvement of the Lisbon city government as well as numerous immigrant associations and entrepreneurs. It is interesting to speculate on why these actors became involved when they did. The literature has shown that carnivals and popular festivals can function to neutralise social conflict; Waterman argues that popular festivals can facilitate marginal groups to express their discontent though symbolic forms, but can also reflect attempts by political and social elites to keep control and maintain modes of spatial division (1998: 60).

The success of the two *Festivals of Diversity* supports the argument that festivals can create a powerful sense of place (Zukin 1991). The hosting of these events in Mouraria amounted to public recognition of the area's cultural diversity, illustrating Waterman's (1998: 55) claim that festivals can transform landscapes and places from everyday settings into temporary environments and contribute to the cultural maintenance of certain groups. Immigrants have indeed consolidated their position in Mouraria, with tourist guidebooks now recommending the neighbourhood as the place to go for a melting pot market experience in the city (Shearaman de Macedo 2001: 166; Hancock 2007: 76).

Eager to capture some of the profits arising from the increased demand for cultural diversity, Chinese entrepreneurs began lobbying for a change in the neighbourhood's identity. In 2000, the President of the China Town Commercial Association tried to transform Martim Moniz Square into the Lisbon Chinatown. In his own words:

> A Chinatown usually attracts tourists, as has happened in New York and Paris. . . . We contracted several Chinese designers to prepare a project to transform the square into a Chinatown (interviews in November 2003 and January 2008, free translation).

The conflict was now out in the open. The trend towards cosmopolitanism and the commodification and marketing of diversity in other cities notwithstanding, Lisbon authorities chose to emphasise the city's Christian origins. The city government's discourse, which sheds light on the complexity of the case, thus juggles two main tenets—one covertly recognises the role of immigrants in stimulating the local market and the poor, aging neighbourhood; the other builds upon historical memory. Discourses around diversity contain further contradictions: while the media portrays Mouraria as the social and economic meeting place of diverse cultures, immigrant entrepreneurs continue to rely on their positions at the low-cost end of the market rather than on their ethnic or cultural characteristics.

New discussions over the urban regeneration of Mouraria began in 2007, with the imagined Chinatown again becoming the focus of politicians, business associations, anti-discrimination and community spokespeople, local entrepreneurs, the media and members of other immigrant communities. The opinion of one city councillor became a lightning rod for the media as it illustrated in stark terms Mouraria's urban identity as advocated by the city government. The episode began when the right-wing city councillor in question declared to a newspaper that Chinese shops should not be permitted in Lisbon's historic quarter because, in her words, 'if they continue to be opened it will never be possible to control small businesses and they will ruin the city's trade'. As an alternative, she suggested Chinese entrepreneurs could convert Martim Moniz Square into a Chinatown.[11]

Voices were raised against the idea, stating that such a policy would be racist, xenophobic and detrimental to the liberal free market. Other critics—recalling that Mouraria had its origins as a segregated area for Moors and Muslims, and that Lisbon Christians around 1506 had killed many of the several thousand Moors who lived there—argued that the creation of ghettos in the city should not be promoted.[12] For their part,

Chinese association leaders made use of neoliberal arguments to point out that it would be unconstitutional for the city government to impede the creation of businesses based on ethnic origin, or to expel Chinese entrepreneurs from certain parts of the city. Others declared that they did not feel discriminated against, and that the creation of a Chinatown in Lisbon would be positive for both Chinese entrepreneurs and local tourism.[13] Attempting to defuse the public discussion a few days later, the mayor of Lisbon stated that the creation of a Chinatown in Lisbon would be outlandish, and guaranteed that the project was not on the city government's agenda nor was it under consideration.

Business associations also took part in the discussions. Some argued that Chinatowns in other cities stimulated local markets. The same associations also argued that instituting a Chinatown would protect native traditional commerce and small shops from Chinese competition.[14] Against such arguments, a left-wing city councillor argued that the expulsion of Chinese shops from the city centre would spell the end of the local market as immigrant entrepreneurs were the cornerstone of the area's economic activity.[15] This leads us to the third discourse.

The Entrepreneurship Discourse

Walking through Mouraria one comes face-to-face with both medieval cultural symbols and immigrant businesses. As Zukin (1995) has argued, cities are built by combining different elements of the space. The third discourse, in responding to the economic interests of the local market, concerns how to combine heritage and multiculturalism. Here members of Mouraria's critical infrastructure define the interactions between the space's medieval characteristics and entrepreneurial dynamism.

As described above, immigrants played a central role in Mouraria's economic revival by bringing new consumers and entrepreneurs to the area. The success of the first settlers expanded the economic opportunities of those who followed. Immigrant investment also contributed to the re-utilisation of abandoned buildings, both for residential and commercial purposes, thus increasing the value of local real estate. Chinese and Indian entrepreneurs in particular were responsible for the gradual transformation of several abandoned apartments into shops, offices and warehouses for import-export activities.

The regeneration of Mouraria included important improvements to its infrastructure. At the end of the 1990s, a car park was built under Martim Moniz Square, the metro station was modernised and public toilets and benches were made available. In the centre of the square, the city government constructed 44 kiosks of no more than four square meters each. Fountains, trees and flowerbeds were also installed, giving the square a fresher and more modern feel, while an adjacent four-star hotel was renovated. Well-served by public transport, Martim Moniz Square has over the years become the core of Mouraria's economic life.

Nevertheless, events in the late 1990s gave some inhabitants and local entrepreneurs the impression that Mouraria was not a safe place to live in or invest.[16] The infrastructure provided by the city council began to attract illegal activities such as drug dealing and mobile phone unblocking, while the police carried out raids in the neighbourhood, arresting many undocumented immigrants. These circumstances prompted the city government to invest more heavily in security measures: cameras were installed, a private security company was contracted and one of the kiosks was converted into a security office.

Several of the shops that had opened in the square's kiosks, including a Lisbon tourist office, were closed. Against this backdrop, a group of Chinese entrepreneurs in 2000 created the commercial association *China Town*, which aimed to increase the supply and demand for Chinese entrepreneurial activities in the district. Seeking to revitalise the area economically, the city government in 2000 agreed with the association's plan to rent all the kiosks to Chinese entrepreneurs, who would then open small retail shops. The above-mentioned plan to transform Martim Moniz Square into a Chinatown was proposed at this juncture. Although the city government supported the association's initiatives to increase entrepreneurial activity in the neighbourhood, its heritage discourse did not allow this development.

Due to a lack of customers, the Chinese association soon decided to stop investing in the kiosks and to re-invest in the Martim Moniz Shopping Centre. To publicise its commitment and to stimulate demand for its businesses, the Chinese association organised a large and successful ad campaign in late 2000.[17]

> After the advertisement campaign the demand for our businesses grew significantly, in terms of both customers and entrepreneurs, namely Portuguese and Roma people (interview with the president of the China Town Commercial Association in January 2008, free translation).

To conclude, the third discourse—which mainly reflects concerns over the economic dynamism of the Mouraria market—demonstrates that the local political authority can implement complementary strategies for both economic and cultural consumption. While immigrant entrepreneurs were important partners in the social and economic revitalisation of the neighbourhood *per se*, the city government manipulated symbolic languages of exclusion to assert its role in defining the city's identity and heritage underlying particular spaces.

CONCLUSION: THE MEETING OF CITY HERITAGE AND THE DIVER-CITY

Mouraria's centuries-long and more recent histories inform the struggle over defining its urban image. Mouraria is at the centre of Lisbon's origins as a city, a disadvantaged area within the modern city, and since the 1980s, a culturally diverse district hosting vibrant immigrant enterprises.

The use and appropriation of public space over time has transformed Mouraria's image. The contemporary neighbourhood reflects a dynamic accumulation of identities, experiences, symbols, populations and cultures. Within the same physical space, as if in a house of mirrors, visitors are transported from a popular, historic Lisbon neighbourhood into a multicultural diver-city. The social construction of Mouraria's image reflects individual appropriation through practices such as giving shops foreign names and new uses of public space; it also reflects city authorities' strategies to define the official perception of the district, its urban identity. These two conflicting rationales—*appropriation* and *perception*—need to be considered when examining the role of immigrant entrepreneurs in the commodification of neighbourhoods.

The case of Mouraria further shows that the classification and manipulation of meanings attached to local spaces is particularly fraught. The discourses—without being contradictory—reflect different rationales, reinforcing the imaginary identity of the city while focusing on entrepreneurship, new users and multicultural imagery. The many images of Mouraria can work in opposition or in cooperation. As Hoelscher (1998: 390) argues, it is within this instability of definitions and interpretations that conflicts arise. In the case of Mouraria, the success of the heritage discourse depends on the ability of its proponents to conceal the multicultural everyday life of the neighbourhood. The tensions between the heritage discourse based on the city's *reconquista* and the role of Christians in the expulsion of Lisbon's Moorish Muslims and the diver-city discourse embracing diverse mother tongues, cuisines and religions is obvious.

This brings to the discussion how the city's heritage, including its symbolic and material patrimony, can condition immigrant entrepreneurial strategies and the advertised identity of spaces. The argument here is that how the historical origins and heritage of a place are interpreted can broaden or narrow immigrant entrepreneurs' opportunities to participate in the re-imaging of districts. Hence one important finding is that the transformation of ethnic neighbourhoods into sites of cultural consumption is much more complex if this conflicts with existing urban or national identities. Cities that rely on history to attract visitors may furthermore have no need for diversity to do the same.

The dialogue between actors within the three identified discourses reflects the existence of a complex and continuous process of interaction to claim ownership and authenticity. Another key finding is that immigrant economic vitality is in itself insufficient to dictate the ethnic identity of a district. As brought to light in the Chinatown episodes, the Chinese entrepreneurs (though implicitly acknowledged) never achieved formal recognition; this thwarted their attempts to commodify the neighbourhood. We see that the political regulatory structure retained its central role, with the power to implement a policy discourse or to inhibit the implementation of competing ones.

Further research is necessary to see whether the trends observed in Lisbon—the limited role for immigrants and the prioritisation of place heritage—can also be seen in cities lacking centuries of material heritage or with longer histories of immigrant settlement. In other words, contextual factors specific to the neighbourhood under study complicate cross-city generalisations on the transformation of immigrant districts into sites for tourism and leisure. As suggested by Appadurai (in Harney 2006: 26), the production of a neighbourhood is not just an exercise of power over a single setting but the countering of several claims to the same space.

This chapter thus examined the subtle strategies of social differentiation contained in the encounters that make up daily life on the streets, on the square and in the shops of Mouraria. Because the symbolic economy is also the result of these micro-level strategies, one possible interpretation is that immigrants will continue to play a key role in Mouraria's imaging and in stimulating urban tourism from below, as Chinese entrepreneurs have already sought to do. Tourism at the grassroots can be promoted by symbols, visual codes and other resources that local people and groups create and use in everyday life (Gotham 2007).

The research on Mouraria highlights how immigrant entrepreneurs' strategies and market dynamics can be contingent on the commodification of space and the renegotiation of urban heritage. In the complex mixture of immigrant populations, tourism, landscape and the heritage myth, Mouraria points to the negotiability and contingency of place-based identities and the challenges faced by districts wishing to become sites of tourism, leisure and cultural diversity.

NOTES

1. The author would like to thank the editors of this book and Claire Healy for their comments on an earlier draft, which greatly aided revision.
2. 'Imaging' here refers to the creative process of defining the image of a place (as used by Vale and Warner 2001).
3. Census data show a steep decline in Mouraria's population, especially after the 1960s. Mouraria had 15,112 residents in 1960; 6,751 in 1991; and only 4,287 in 2001. This was accompanied by an increase in available housing: the vacancy rate was 7 per cent in 1992.
4. For further details see Oliveira (2007) and Oliveira and Costa (2008).
5. Between 1981 and 2001, there was a 52.7 per cent decline in the number of native entrepreneurs. Over the same period, there was a 118.8 per cent growth in the number of foreign entrepreneurs.
6. Martim Moniz, a warrior killed by the Moors during the conquest of Lisbon, is known for his role in ensuring Christian victory. When the Moors opened the fortress gate, Moniz used his own body to keep it open, allowing the Christians to enter the city. King Afonso I proclaimed the city gate would henceforth be known as Martim Moniz.
7. The President of Socorro, the parish in which Mouraria is located, recently told the media that of its 15,000 residents, 11,000 are immigrants, while Muslims are now the dominant religious group (*Lusa*, 3 April 2008).
8. See Malheiros (2005) and several newspaper articles in *Público*, 16 March 2003: 30; *Diário de Notícias*, 2003: 18; *Revista Focus,* 5 September 2007: 34–39.
9. Lisbon's *Multicultural and Cosmopolitan Policy* was defined in 1993. Under this policy the city government created the Municipal Board for Immigrant Communities and Ethnic Minorities [Concelho Municipal das Comunidades Imigrantes e das Minorias Étnicas].

10. See Mapril (2001) and Menezes (2004).
11. For further details see her interview in *Expresso*, 8 September 2007: 1–9.
12. For further details see *Público*, 18 September 2007: 44.
13. For further details see articles in *Gobal*, 14 September 2007: 4; *Diário de Notícias*, 14 September 2007: 30; *Notícias da Manhã*, 20 September 2007: 1–3.
14. The vice president of the Lisbon Traders Association declared that if the Chinese were concentrated in one place it would be better for small Portuguese shops throughout the city (cited in newspaper article in *Notícias da Manhã*, 20 September 2007: 1–3). For more on this argument see several articles in *Diário Económico*, 12 September 2007: 36–37 and *Diário de Notícias*, 13 September 2007: 1–10.
15. For further details on this city councillor's declaration, see *Jornal de Negócios*, 12 September 2007: 35 and 40.
16. In 1999 the managers of the four-star hotel in Martim Moniz Square organised a petition with other local entrepreneurs to lobby the city government to guarantee safety in the neighbourhood.
17. The President of the China Town Commercial Association stated in a January 2008 interview that around 25 million escudos (about €125,000) was spent, and the campaign was paid for by entrepreneurs.

REFERENCES

Almeida, M.R. (1998) *Vitimização e insegurança no concelho de Lisboa*. Lisbon: Gabinete de Estudos e Planeamento do Ministério da Justiça.

Collins, J. (2007) 'Ethnic precincts as contradictory tourist spaces', in Rath, J. (ed.) *Tourism, Ethnic Diversity and the City*. New York and London: Routledge, 67–86.

Costa, F. (2007). "Baixa sem Lojas Chinesas", *Expresso*, 8 September 2007: 1–9.

Esteves, A. (1998) *A criminalidade na cidade de Lisboa*. Lisbon: Edições Colibri.

Gotham, K.F. (2007) *Tourism, Culture, and Race in the Big Easy*. New York: NYU Press.

Felner, R. (2003), "Martin moniz aprumou-se para receber Jorge Sampaio", in *Público*, 16 March 2003: 30.

Grouveia, B. (2007), "Chinatown turística", in *Notícias da Manhã*, 20 September 2007: 1–3.

Hall, M. and Rath, J. (2007) 'Tourism, migration and place advantage in the global cultural economy', in Rath, J. (ed.) *Tourism, Ethnic Diversity and the City*. New York and London: Routledge, 1–24.

Halter, M. (2000) *Shopping for Identity: The Marketing of Ethnicity*. New York: Schocken Books.

Handcock, M. (2007) *Rough Guide Directions Lisbon*. London: Rough Guides Ltd.

Harney, N. (2006) 'The politics of urban space: Modes of place-making by Italians in Toronto's neighbourhoods', *Modern Italy*, 11(1): 25–42.

Harvey, D. (2003) 'National identities and the politics of ancient heritage: Continuity and change at ancient monuments in Britain and Ireland, c.1675–1850', *Transactions of the Institute of British Geographers*, 28(4): 473–87.

Hoelscher, S. (1998) 'Tourism, ethnic memory and the other-directed place', *Cultural Geographies*, 5: 369–98.

Jarman, N. (1997) *Material Conflicts: Parades and Visual Displays in Northern Ireland*. Oxford: Berg.

Johnson, N.C. (1999) 'Framing the past: Time, space and the politics of heritage tourism in Ireland', *Political Geography*, 18: 187–207.

Kloosterman, R. and Rath, J. (2001) 'Immigrant entrepreneurs in advanced economies: Mixed embeddedness further explored', *Journal of Ethnic and Migration*

Studies, Kloosterman, R. and Rath, J. (eds) *Special Issue on 'Immigrant Entrepreneurship* 27(2): 189–201.
LUSA (2008), "Freguesia do Socorro conta com 11 mil imigrantes", *LUSA*, 3 April 2008.
Malheiros, J.M. (2005) 'Urban xenophilia: Journeys through a multicoloured city', in Brasil, D. and Galvão Lucas, M. (eds) *Em trânsito—Mobilidade e Vida Urbana*. Lisbon: Goethe-Institut Lisbon, 309–17.
Mapril, J. (2001) *Chineses no Martim Moniz*. SociNova Working Papers Series 19.
Menezes, M. (2004) *Mouraria, Retalhos de um imaginário. Significados Urbanos de um bairro de Lisboa*. Oeiras: Celta.
Morais, F. (2007), "Chinese aprovam 'Chinatown' em Lisboa", in *Diário de notícais*, 14 September 2007: 30.
———. (2007), "Chinatown ajudara negocios e turismo", in *Global*, 14 September 2007: 4.
Moura, L. (2007), "Chinatown", in *Jornal de Nargócios*, 12 September 2007: 35.
Oliveira, C.R. (2007) 'Understanding the diversity of immigrant entrepreneurial strategies', in Dana, L-P. (ed.) *Handbook of Research on Ethnic Minority Entrepreneurship*. Cheltenham and Northampton: Edward Elgar, 61–82.
Oliveira, C.R. and Costa, F.L. (2008) 'Being your own boss: Entrepreneurship as a lever for migration?', in Fonseca, L. (ed.) *Cities in Movement: Migrants and Urban Change*. Lisbon: Centro de Estudos Geográficos, 241–66.
Pinto, M. (2007), "Chinatown ou cidade aberta?", in *Jornal de Negócios*, 12 September 2007: 40.
Rath, J. (2007a) *The Transformation of Ethnic Neighbourhoods into Places of Leisure and Consumption*. CCIS Working Paper 144, University of California, San Diego.
———. (ed.) (2007b) *Tourism, Ethnic Diversity and the City*. London and New York: Routledge.
Shaw, S., Bagwell, S. and Karmowska, J. (2004) 'Ethnoscapes as spectacle: Reimaging multicultural districts as new destinations for leisure and tourism consumption', *Urban Studies*, 41(10): 1983–2000.
Shearaman de Macedo, G. (2001) *Time Out Lisbon*. London: Penguin Books.
Sobral, F. (2007), "Chinatown ou cidade aberta?", in *Jornal de Negócios*, 12 September 2007:40.
Tavares, R. (2007), "Chinatown em Lisboa divide costa de Sá Fernandes", in *Diário Económico*, 12 September 2007: 36–37.
———. (2007), "Salazasinha Nogueira Pinto", in *Público*, 18 September 2007: 44.
Teixeira, R. (2007), "Os novos portugueses", in *Revista Focus*, 5 September 2007: 34–39.
Tilley, C. (2006) 'Introduction: identity, place, landscape and heritage', *Journal of Material Culture*, 11: 7–32.
Vale, L. and Warner, S.B. (eds) (2001) Imaging the City: Continuing Struggles and New Directions. New Brunswick: Center for Urban Policy Research.
Waterman, S. (1998) 'Carnivals for elites? The cultural politics of arts festivals', *Progress in Human Geography*, 22(1): 54–77.
Wells, K. (2007) 'The material and visual cultures of cities', *Space and Culture*, 10: 136–144.
Yeoh, B. (2005) 'The global cultural city? Spatial imagineering and politics in the (multi)cultural marketplaces of South-east Asia', *Urban Studies*, 42(5/6): 945–58.
Zukin, S. (1995) *The Cultures of Cities*. New York: Blackwell.
———. (1991) *Landscapes of Power: From Detroit to Disney World*. Berkeley: University of California Press.
———. (1987) 'Gentrification: Culture and capital in the urban core', *Annual Review of Sociology*, 13: 129–47.

7 Symbols of Ethnicity in a Multi-ethnic Precinct

Marketing Perth's Northbridge for Cultural Consumption[1]

Kirrily Jordan and Jock Collins

INTRODUCTION

For over a century, waves of immigrants have left their mark on Northbridge, just across the tracks from Perth's main railway station. Chinese, Italian, Greek, Vietnamese and many other new arrivals settled there, often moving on over time. Chinese immigrants in the late 1800s were followed by Greeks and Italians in the mid-twentieth century and a more diverse array of immigrants since the 1970s. Many set up their own small retail and hospitality businesses along Northbridge's main streets. They also established places of worship and ethnic associations that feature prominently in the area's built and social environments. Today, Northbridge is home to a great variety of immigrant enterprises, reflecting the multi-ethnic character of Northbridge itself.

This chapter examines the development of Northbridge as an ethnic neighbourhood, or precinct, over more than a hundred years of (changing) immigrant settlement. Over this period, Northbridge has had a number of ethnic faces, and its identity remains in flux as a recent attempt to rebrand the area as Perth's *Chinatown* failed. We draw on fieldwork with key stakeholders in the development of Northbridge as an ethnic precinct (members of local government and ethnic community organisations, town planners, ethnic entrepreneurs) to explore the complex and sometimes contradictory interactions between them in shaping Northbridge's identity. Keen to capitalise on the area's ethnic diversity to attract investment and tourism, both local and state governments have engaged in a 're-visioning' of Northbridge to highlight its distinctively ethnic character.

Our contribution explores the various attempts to market Northbridge's ethnic diversity. It first outlines the history of the area, detailing the changing patterns of immigrant settlement and the establishment of ethnic enterprises. It then examines recent attempts to rebrand part of Northbridge as a *Chinatown*, outlining the neighbourhood's complex institutional environment and the roles of various actors within current redevelopment strategies. We argue that Northbridge's evolving identity as an ethnic precinct—and issues of 'safety' in the neighbourhood—are driven by interaction between key stakeholders within the critical infrastructure, regulators, immigrant entrepreneurs and ethnic community representatives.

NORTHBRIDGE, PERTH, WESTERN AUSTRALIA

Northbridge is an ethnic neighbourhood and restaurant precinct in the heart of Perth, the capital city of Western Australia (WA) (see Figure 7.1). Only a few hundred metres from Perth's central business district, it is separated from the city by the main railway line. The separation from the city is not new: from its earliest days, Northbridge was isolated by its swampy terrain and was important for local Aboriginal people (*Nyoongars*) displaced from more accessible land to the south. The area also attracted others among the city's poorer residents, including Chinese immigrants (Gregory 2007: 3). Through the late 1800s it became Perth's 'red light' district, housing brothels, Chinese gambling houses and opium dens. Divided from the main city by the railway line in 1881, in the minds of many Perth residents Northbridge was definitely on the 'wrong side of the tracks'.

From the 1890s, Chinese immigrants in Northbridge began establishing their own businesses, including laundries, market gardens and furniture factories (King 1998; Peters 2007). Chinese immigrants were at the time

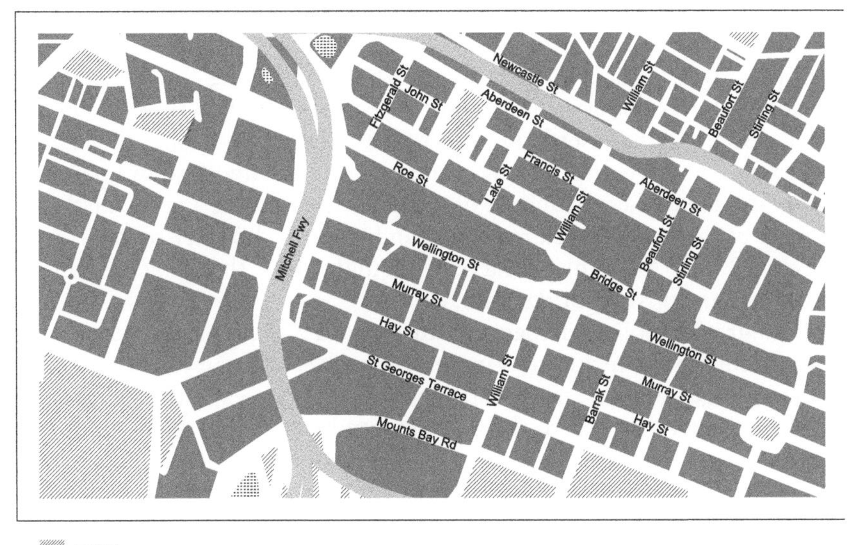

Figure 7.1 Map of Northbridge. *Map designed by Ana Sala.*

a racialised minority in Australia; entrepreneurship was in part a response to their 'blocked mobility' in the labour market (Collins 2002). Because of the 'white Australia policy' introduced in 1901, the Chinese population in Australia fell from just under 30,000 in 1901 to 9,000 in 1947 (Choi 1975: 27). A small number of Chinese remained in Perth. Northbridge became their central meeting place, with the Chung Wah Association, established in Northbridge in 1909, active in promoting the rights of local Chinese residents and entrepreneurs (King 1998). With its concentration of Chinese businesses and social activities, Northbridge soon gained the reputation of being Perth's 'China Town' (Peters 2007: 1).

At the end of the First World War, Greek and Italian immigrants began arriving in Northbridge in significant numbers. They were seen by their host country as the 'Chinese of Europe', racialised minorities who could replace the Chinese in the least desirable jobs (de Lepervanche 1975; see also Church 2005; Price 1963). Faced with institutional and popular racist hostility, many Greek and Italian immigrants in Australia found work through self-employment; some established small businesses—particularly cafés, green grocers and delicatessens—in Northbridge (EPRA 2000a; Yiannakis 1996). The Greek and Italian cafés were the first to be allowed *al fresco* dining, now a Northbridge institution and an important part of its contemporary appeal.

Like the Chinese, most of the early southern European immigrants were men, and illegal gambling houses soon replaced the opium dens. Hostels provided housing and functioned as unofficial welfare and employment agencies. The number of Italian and Greek immigrants in Northbridge increased considerably after the Second World War (Kringas 2001; Stransky 2001) via 'chain migration' (Price 1963), often paid for by family members already settled in Northbridge. While the Chung Wah Association remained, most Chinese shops and residents had by this time left Northbridge. Some Chinese settled in other parts of Perth. No longer referred to as Perth's China Town, the concentration of Greek and Italian homes and businesses led to the new names of 'Little Megisti' (reflecting the birthplace of most of the area's Greek residents) and 'Little Italy' (City of Perth 1989: 5; Peters 2007). After this string of colloquial names, the whole district was officially named Northbridge in 1981.

Immigration policy changed again in the 1970s. Southeast and South Asian immigrants, including Vietnamese refugees, began to arrive in Australia in large numbers, many settling in Northbridge and the surrounding suburbs. Like those before them, many established their own small businesses. Some of the area's Greek and Italian shops remained but were now joined by Vietnamese and Chinese restaurants, Asian butchers, and a number of professional services targeting Asian clienteles. The growth of Vietnamese shops and restaurants along the north of William Street prompted the new nickname 'Little Saigon' (Peters 2007).

The composition of Northbridge's population continues to evolve. The recent real estate boom in Perth and the proximity of Northbridge to the city led to significant increases in housing costs and the gentrification of existing properties. Wealthy urban professionals displaced many of the earlier immigrants, with around 64 per cent of Northbridge residents being born in Australia or the UK. The largest non-Anglo-Celtic immigrant groups come from Malaysia (8 per cent of the immigrant population), followed by Indonesia (7.5 per cent), Hong Kong (7 per cent), Thailand and Vietnam (both 6.5 per cent).[2] Only 3 per cent of immigrants in 2006 were born in Italy, while none of Northbridge's residents were born in Greece (ABS 2007). As the residential population of Northbridge changed, many immigrant entrepreneurs rebranded their products to appeal to the professional middle class. Several long-standing immigrant grocery stores now market themselves as gourmet food stores, selling high quality products and a wider range of European, Asian and Middle Eastern foods. With its diverse restaurants and food stores, Northbridge has become one of the most popular sites in Perth for locals and tourists alike to experience a range of ethnic cuisines.

Today, the built environment of Northbridge, its streetscape and landscape, is clearly multi-ethnic. As well as being visible in its small businesses, the traditional role of Northbridge as immigrant repository is evident in its community buildings, many of which are now important icons. The area is home to Perth's first Greek Orthodox Church, first Italian Club, first Vietnamese Buddhist Temple and Western Australia's first mosque. In addition, it has remained an important meeting place for the Nyoongar community and a key site for Nyoongar political organisation.

NORTHBRIDGE AND THE TOURISM TRADE

The growth of cultural industries and tourism is an important characteristic of modern developed economies like Australia. These industries intersect in the field of *cultural tourism*. Cultural tourism is recognised as an important agent of economic and social change in Europe (Richards 2005: 10). It includes tourism to traditional cultural attractions such as museums, art galleries and opera, but also incorporates new forms of tourism associated with cultural activities. Particular areas for further research include tourism featuring ethnically diverse populations in cosmopolitan cities as well as cultural attractions related to the ethnic diversity that has accompanied immigration to countries across the Western world (Rath 2007), including Australia (Collins and Kunz 2007). We refer to this as *ethnic cultural tourism*. As Hall and Rath (2007: 6) remind us, 'Tourism, migration, ethnic diversity and place are structurally related in many ways, but this interrelationship has been surprisingly understudied'. We are particularly interested in one subset of ethnic cultural tourism: tourism in ethnic precincts in cosmopolitan cities.

While Little Megisti, Little Italy and Little Saigon were unofficial names, they illustrate how Northbridge by the early 1900s had begun taking on the characteristics of an ethnic precinct. A key feature of ethnic precincts is the provision of ethnic food and restaurants (Gabaccia 1998; Warde 1997; Warde and Martens 2000). The majority of immigrant entrepreneurs in Northbridge established restaurants or wholesale and retail food stores, a common pattern among immigrant entrepreneurs reflecting the relatively low entry costs and qualifications needed in these industries (Collins *et al.* 1995). While most of Northbridge's early immigrant entrepreneurs catered to the local market, in recent decades the restaurants and food stores in particular have become popular among tourists and visitors. This reflects changing public attitudes to immigrant minorities and ethnic foods; the federal policy shift towards multiculturalism in the 1970s was accompanied by the increasing willingness of locally born consumers to try new and 'exotic' goods (see for example Fitzgerald 1997).

Ethnic precincts often host ethnic community organisations and activities. Northbridge hosts several annual festivals, some reflecting the area's ethnic history. Chinese and Vietnamese New Year celebrations, the Greek *Glendi* festival and a world music festival are all held in Northbridge each year. The annual *Northbridge Festival* highlights contemporary arts and acknowledges Northbridge's ethnic diversity; the 2007 programme included a Chinese orchestra and history tours pointing out important sites in the area's immigrant heritage. Northbridge is also a key site for events during the Perth International Arts Festival and the annual gay and lesbian *Pride Parade.*

Licensing laws have altered Northbridge's character and influenced its transformation into a destination for tourists and visitors—an example of how regulation regimes help shape the tourist experience (Hoffman et al. 2003). Northbridge is one of the few districts in Perth where clubs and pubs remain open into the early morning. It has thus become synonymous with late-night entertainment and one of the key areas in Perth for backpackers' hostels, which draw thousands of tourists to the area annually. Northbridge also remains Perth's 'red light' district, with its main streets housing a number of sex shops and massage parlours. This has mixed effects. While it attracts visitors to Northbridge, others prefer to eat out in nearby suburbs with a more 'family' atmosphere.

Tourism in Northbridge has also developed around more traditional cultural attractions. In particular, the Perth Cultural Centre on Northbridge's east side houses the Western Australian Museum, Art Gallery of Western Australia and Perth Institute of Contemporary Arts. Tourist brochures for Perth advertise Northbridge's mainstream cultural institutions, late-night entertainment and ethnic diversity as a 'package' of attractions, encouraging tourists staying elsewhere in Perth to visit the area during their trip. While licensing laws, the development of cultural institutions and immigrant entrepreneurs' re-orienting to the mainstream market have all opened up opportunities for tourism in Northbridge, little is known about its customer base. A 1989 study found that many shoppers came to the area due to the availability of 'ethnic goods' (MEAC and OMA 1989: 31).

TOURISM IN NORTHBRIDGE AND THE REGULATORY REGIME

Ethnic precincts do not exist in institutional vacuums. They are shaped by the interaction of producers (immigrant entrepreneurs), consumers, governments, community leaders and members of what Zukin (1995) calls the 'critical infrastructure' (including tourism and food critics and place marketers). For example, local or city governments may play key roles in naming, marketing and investing in ethnic precincts as well as in shaping public façades (see for example Anderson 1990; Ip 2005; Pang and Rath 2007). Ethnic precincts' regulatory regimes can best be understood through the lenses of two distinct theoretical models: the 'mixed embeddedness' approach to understanding immigrant entrepreneurship (Kloosterman and Rath 2001) and the framework for understanding urban tourism known as 'regulation theory' (Hoffman et al. 2003).

Ethnic and immigrant entrepreneurship has been the subject of much academic research (Collins *et al.* 1995; Kloosterman and Rath 2003; Light and Rosenstein 1995; Light and Gold 2000; Rath 2000; Waldinger *et al.* 1990). Ethnic entrepreneurs here are central to the creation of ethnic precincts, as well as to their long-term vitality (Collins *et al.* 1995; Collins 2005). The mixed embeddedness approach further recognises that immigrant enterprises are embedded in the economic, social and political structures of broader society. For example, it recognises the complex interplay of entrepreneurs' social networks, local and national policies on immigration and business ownership and variations in the market dynamics for different types of goods and services as key factors in the shaping of opportunities and limitations for immigrant entrepreneurship (Kloosterman and Rath 2001; Rath 2002). The regulatory environment for immigrant entrepreneurship can have a profound effect on the formation and character of ethnic precincts. Where opportunities for entrepreneurship are limited, ethnic precincts are less likely to emerge. Strategies for urban renewal may include the development of new ethnic precincts or the revitalisation of existing ethnic neighbourhoods through streetscape design or the promotion of ethnic festivals—popular strategies for city governments given urban economies' increasing dependence on culture industries and the intensified competition to attract both tourists and mobile financial capital (Lash and Urry 1994; Zukin 1995).

In Perth, the local (City of Perth) and state governments embarked on a programme to attract investors and consumers to the city in the 1990s. While Western Australia's economy experienced a short boom in the 1980s, it was in recession by the early 1990s, with much of Perth's office space lying empty (Iveson 2000). Both local and state governments saw the solution in creating a new identity for Perth's central area, including Northbridge. As Iveson (2000: 229) describes, 'one of the main roles identified for government was investment in public space improvements which it was hoped would create a "sense of place" and thus encourage shoppers, tourists, residents and employers back to the Central Area'.

In Northbridge, the programme of action has been concerted and ongoing. Redevelopment has been characterised by a high degree of cooperation between state and local government agencies. In 2000 the state government agency East Perth Redevelopment Authority (EPRA) was given control over a large portion of Northbridge, previously governed by the City of Perth. The entry of EPRA divided 'greater' Northbridge[3] into three separate local jurisdictions: roughly, the City of Perth between Roe and Aberdeen Streets, EPRA along both sides of Newcastle Street, and the Town of Vincent north of Newcastle Street to Brisbane Street.

PERTH

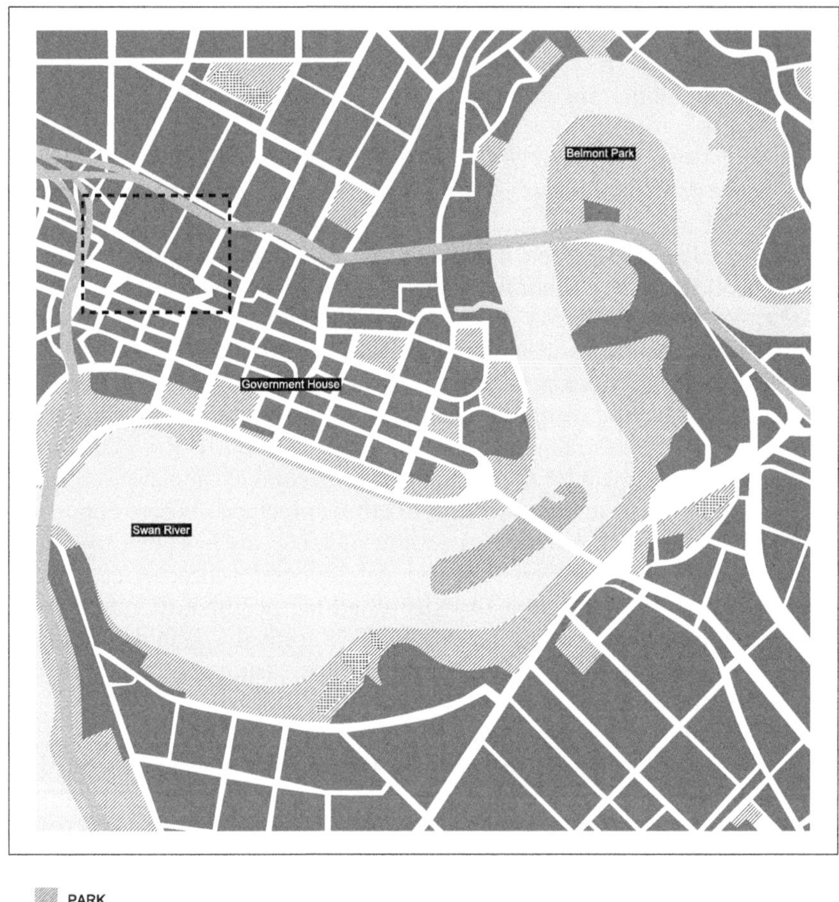

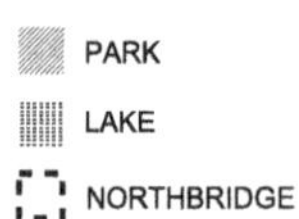

Figure 7.2 Location of Northbridge in Perth. *Map designed by Ana Sala.*

The continued revitalisation of Northbridge is co-ordinated under the Northbridge Board of Management, established in 2005. Chaired by a City of Perth councillor, the board includes representatives from the City of Perth, EPRA, the Western Australian Police, the state Department of Premier and Cabinet (DPC) and Tourism Western Australia. It also includes members of the Business Improvement Group of Northbridge (BigN). The Group, established in 2002 on the recommendation of the City of Perth, includes local entrepreneurs, residents and community groups; its purpose is to improve physical, aesthetic and community amenities for both locals and visitors. The recommendations of the Northbridge Board of Management are put into action by a full-time precinct manager. Reflecting the close co-ordination of the actors involved, the precinct manager is funded by the City of Perth, EPRA and the DPC; she also acts as the full-time representative of the BigN, liaising with local business owners.

The strategy for creating a 'sense of place' in Northbridge has included major physical redevelopments such as the construction of several multistory commercial and residential complexes as well as social housing and retail space. EPRA bought several retail properties on William Street and is developing a rental plan to dictate the types of businesses that can operate there. The City of Perth is planning to develop the 'Northbridge Piazza', a public square in central Northbridge, including space for commercial tenants, community facilities and a gallery, as well as an open area, stage and 24-hour LED screen for performances and special events. The aim is to make the area a 'commercial, cultural and creative hub' in the city (City of Perth 2007). The plans to create a sense of place to attract consumers to Northbridge include two further elements: a focus on the area's ethnic diversity as well as strategies to improve safety. These issues will be addressed in turn below.

'SELLING' NORTHBRIDGE AS AN ETHNIC PRECINCT: MULTI-ETHNIC OR A CHINATOWN?

In describing the transformation of cities into centres of cultural consumption, Zukin (1995) emphasises the importance of creating a 'vision' for the city to capture the imagination of tourists and corporate elites. She notes that this is a contested process with 'constant political pressures by interest groups and complex interwoven networks of community groups, corporations, and public officials' generating multiple visions that may not be easily reconciled (Zukin 1995: 14). These contests over representation are open-ended and ongoing. In Northbridge, the 'vision' has included a focus on the area's ethnic diversity in at least two distinct ways. The first includes two attempts to market the area as Perth's 'Chinatown'. The second includes a number of approaches drawing on Northbridge's multi-ethnic character.

In the early 1980s, with the influx of Asian immigrants into the area, a small group of Malaysian Chinese businesspeople began planning a Chinatown in Northbridge. Their plan aimed to 'retain and enhance the character and ethos of the Orient within a totally planned modern environment', complete with Chinese arches, awnings and iconography (*Chinatown* 1982: 1). The proposal roughly coincided with the development of formal Chinatowns in Sydney and Melbourne on Australia's east coast (see Anderson 1990). All three were attempts to market an area on the basis of its ethnic character and were indicative of similar developments in other Western cities. However, the motivation for developing a Chinatown in Northbridge had a unique aspect: in promoting the concept, the developers explicitly referred to its role in securing permanent residency visas for foreign nationals who would be employed there (*Chinatown* 1982).

The developers purchased land on Northbridge's southern edge and constructed Chinese arches leading to two parallel laneways. The original plan was to purchase surrounding properties to allow over 30 specialist Asian food stalls, a Chinese theatre, a 49-bed hotel, entertainment facilities and offices (*Chinatown* 1982: 1). This broad-ranging plan, however, was never fully implemented due to three key factors: poor positioning, the reluctance of property owners to sell to the developers and the late 1980s blow-out in interest rates. The development was thus limited to the two short laneways lined on either side by narrow shops, with Chinese arches marking the entry to the laneways at one end.

The small 'Chinatown' met with limited success. The two laneways run along a strip of land between a busy main road and railway line at one end (Roe Street) and a parking lot at the other. While the area is close to the main shopping streets of Northbridge, particularly James and William Streets, there is no indication on either that Chinatown is around the corner. While the arches on Roe Street provide a visual cue, they are of limited use in attracting visitors as very few pedestrians walk along this street, an area dominated by large industrial and commercial buildings and virtually devoid of retail premises. Today the shops along Chinatown's two laneways include a handful of Chinese restaurants, a hairdresser and a Chinese tea house. A number of shops remain vacant.

While the plans for this first Chinatown were never fully realised, 20 years later a second attempt to create a formal 'China Town' in Northbridge began to gather momentum. The attempt was spearheaded by Town of Vincent Mayor Nick Catania, who wanted to develop a 'readily identifiable and popular precinct' that 'for want of a more appropriate name might be called "China Town"'. He suggested that a formally defined and marketed Chinatown was appropriate for any city wanting to be taken seriously on the global stage, with 'almost every major city in the world [boasting] a colourful and culturally diverse "China Town"' (Town of Vincent 2006: 1).

Consultations with business owners and residents on upper William Street, however, made it clear that the proposal had limited support. In particular, many of the businesses were owned by non-Chinese entrepreneurs, including Vietnamese, Thai and Japanese. Even some ethnic Chinese entrepreneurs felt little connection with China, having been born in Southeast Asia. The proposal was also opposed by some members of Perth's leading Chinese organisation, the Chung Wah Association, who felt that establishing a formal Chinatown would alienate members of the non-Chinese community. Their concerns reflected a common academic critique of ethnic precincts: that an area's ethnic 'branding', while presumed to appeal to tourists and visitors, can represent ethnicity in a way that the local ethnic community feels is inappropriate (Anderson 1990; Lin 1998; Collins and Kunz 2005) or that reinforces notions of the 'ethnic Other' (Hage 1997). When 'branding' ethnic precincts, decisions over which symbols are appropriate, who decides, and how, are thus crucially important. Acknowledging community concerns in Northbridge, Nick Catania backed down on the plans for a China Town:

> In deference to the area's diversity of cultures, both past and present, we refer to the proposal [for redevelopment] simply as William Street. . . . Whilst the name may be a bit contentious, the philosophy behind the area is not (Town of Vincent 2006: 1).

With the plan to create a formal Chinatown sidelined, the redevelopment of William Street will include public art that incorporates symbols from the area's numerous ethnic groups.

The second and more successful approach to marketing Northbridge's ethnic diversity has involved greater recognition of its multi-ethnic character. Broadly speaking, this has been done in two ways. Drawing on input from public forums and representatives of government, business groups and Greek, Italian, Chinese and Indigenous community organisations,[4] EPRA constructed public places that recognise some of the area's diverse ethnic groups. For example, 'Piazza Nanni' is a public square that celebrates the first Italian priest to oversee Northbridge's Italian Catholic congregation, while 'Plateia Hellas' is a square that acknowledges Northbridge's Greek community (EPRA 2000b). Calling on public suggestions for street names that recognise prominent Northbridge figures, EPRA also named several roads in honour of local Italian, Greek and Chinese families. These include Via Torre, Kakulas Crescent and Hoy Poy Street.

Taking a somewhat different approach, the state Department of Premier and Cabinet has been funding the Northbridge History Project. The project aims to recognise the area's history through public signage, exhibitions and walking tours and to use this heritage to boost business and tourism (Northbridge History Project 2007). The Northbridge History Project has involved extensive community consultations over several years. A new initiative, the 'Look Up! Northbridge' history tours, introduces visitors to some of the key ethnic institutions in the area.

While much of the recent effort to market Northbridge's ethnic diversity to tourists and visitors has drawn on community consultation, criticisms remain. Some long-standing Anglo-Australian residents have argued that in the effort to revalorise Northbridge's immigrant history, the area's Anglo heritage has been overlooked. In addition, while the development of Plateia Hellas drew on input from public forums and ethnic community representatives, there remains disagreement over whether its location on Lake Street is appropriate: while the street was home to several Greek residences, it also housed a number of Italian shops. This Italian heritage is not recognised in the Plateia.

Such concerns have not led to widespread dispute or resentment, with most parties willing to accept the outcomes of the planning process. The concerns do, however, reflect the difficulties of achieving consensus over symbols of ethnicity in areas with multi-ethnic histories. These relate to the contradictions of authenticity and legitimacy at the interface of tourism and ethnicity (Collins 2007: 74–79). Even when ethnic precincts take on the identity of a single ethnic group, such as the Chinese in Chinatown, the legitimacy problem becomes one of which among the 100 ethnic Chinese community groups in Sydney are to be included in decision-making. This can have a lasting impact on how ethnic identity is represented in the built environment and in marketing materials. For multi-ethnic precincts like Northbridge or Sydney's Cabramatta, the problem of legitimacy is readily apparent. Not only are there a range of voices but, as Meethan (2001: 27) argues, the symbols themselves are 'multivocal, that is, they have the capacity to carry a range of different, if not ambiguous and contradictory meanings'.

Concerns over the use of ethnic symbols may also reflect the limitations of traditional planning processes such as public forums and key stakeholder consultations in engaging people from ethnic minority groups, effectively limiting the number and range of voices heard (Stewart *et al.* 2003; see also Burayidi 2003; Sandercock and Kliger 1998a, 1998b; Thompson 2003). These problems are compounded where ethnic minority groups are multiple-origin, small or scattered, a condition described by Vertovec (2007) as 'super-diversity'. The political implications are significant. Where different ethnic groups, organisations or even individuals have differential access and input into decision-making, it reflects cultural unevenness in urban citizenship, with some denied the full accordance of rights and responsibilities in decision-making over the design and use of public space (Dunn 2003: 154).

INCLUSION AND EXCLUSION: SAFETY IN THE ETHNIC PRECINCT

'Selling' Northbridge to tourists and visitors has also involved efforts to improve safety. The issue of tourist safety is central to any government tourist strategy. Few people would want to go to a place where their safety or their family's safety is at risk. Control and surveillance therefore play an integral part in the development of tourism in general, especially in potential tourist precincts such as ethnic neighbourhoods. Body-Gendrot (2003: 39) emphasises the 'techniques of social control and security' required by mega event tourism, such as the Olympic Games or World Cup soccer, while Judd (2003: 23) points out that building tourist places as *fortress spaces* is one response to managing issues of tourist safety. Borrowing from Foucault, Edensor (1998, 2001) notes that in shopping malls there is a 'remorseless surveillance through panopticon visual monitoring'. Shopping is encouraged, but, as Judd (2003: 29) argues, 'aimless loitering is discouraged or forbidden'.

A number of aspects of control and surveillance relate to ethnic precincts. The first relates to the historical construction of minority immigrant and Indigenous communities as criminal (Collins *et al.* 2000), so that the places and spaces where they concentrate acquire a criminal reputation. This is reinforced by the way racism, prejudice and xenophobia construct immigrant and Indigenous minorities as the criminal 'Other' threatening the safety of the host society (Poynting *et al.* 2004). Ethnic precincts are thus constructed as places of gambling, drugs, prostitution and criminal gangs at the same moment that they become exotic places. Ironically, this criminal feel can also be an attraction to tourists. Chinatowns the world over have always had a criminal aspect, at least in popular perception. According to Kinkead (1993: 47), in the first decades of the twentieth century tourists 'went to Chinatown to ogle vice: guidebooks warned of the immorality and filth of the quarters. The sightseers hired guides to show them opium dens, slave girls, and sites of lurid tong murders. Bohemians visited to smoke opium and drift away into hazy dreams.'

In Northbridge, safety at night has been an ongoing concern over the last two decades. Media reports have linked crime to the area's ethnic diversity, focusing in particular on crime purportedly committed by Nyoongars and alleged Vietnamese gangs (Mac Arthur 2007; MEAC and OMA 1989; Rayner 2003). While they exaggerate the link between crime and ethnicity, concerns about violence in Northbridge have not been unwarranted. Being one of Perth's main venues for late-night entertainment, alcohol-fuelled violence is a recurrent problem. Late-night brawls outside Northbridge's pubs and clubs are not uncommon, although the perpetrators are not limited to any one ethnic group and the attacks are rarely racially motivated. However, the tendency of some Nyoongars to congregate and drink in Northbridge's parks or walk along Northbridge's main streets asking restaurant patrons for money has exacerbated perceptions of a lack of safety.

In response to concerns about personal safety and the potential impact on business, the Northbridge Board of Management has made safety a key issue. To limit the outbreak of fights outside pubs and clubs, regular police patrols are more visible in Northbridge under the revitalisation programme. Police work alongside the Nyoongar Patrol, introduced in 1998 in response to concerns about Indigenous people congregating in Northbridge's public spaces. In contrast to trends in places such as the United States that have witnessed the increasing privatisation of security (see for example Zukin 1995), the Nyoongar Patrol is a publicly-funded, community-based programme where Aboriginal workers with limited formal powers patrol Northbridge and try to resolve situations before police action is needed. The aim is to prevent enmeshment in the criminal justice system (Blagg and Valuri 2004; Government of Western Australia 2008). The Nyoongar Patrol, praised by police and other government agencies as a critical early intervention and outreach strategy, has been effective in reducing public complaints (Government of Western Australia 2008). The presence of young Indigenous people on Northbridge's streets is also controlled through a youth curfew, introduced by the state government in 2003. Under the curfew any unaccompanied children in Northbridge at night can be picked up by police and taken back to their families or to a refuge (Cox 2003).

These approaches to policing in Northbridge highlight the tensions between everyone's right to use public space, including young people, and the negative impact that a history or reputation of conflict, violence or 'public nuisance' has on tourism and the profitability of ethnic enterprises. Entrepreneurs in Northbridge, particularly those in hospitality businesses that rely on evening trade, have long expressed concern that they are losing business due to the reputation of the area as unsafe (there is no evidence that Northbridge attracts the kinds of 'vice tourists' identified by Kinkead). While Northbridge's entrepreneurs may therefore applaud the youth curfew and its potential to reduce public nuisance, it has been criticised by lawyers and rights activists for its disproportionate impact on Indigenous youth, with almost 90 per cent of children picked up in the first three months of the curfew being Aboriginal (Rayner 2003). It has also been argued that the curfew reinforces false stereotypes of young and Indigenous people as being disproportionately responsible for crime, while failing to address the underlying causes of young people being out alone at night (Cox 2003; Mac Arthur 2007). In addition, while the state government and police have praised the work of the Nyoongar Patrol, it has been criticised by local business owners who see it as a 'taxi service' taking intoxicated people to sobering centres rather than deterring criminal behaviour (Government of Western Australia 2008). The City of Perth has recently withdrawn much of its funding for the programme, backing business concerns that it does not act as a deterrent to criminal conduct (*The West Australian* 2005).

The state government's continued support of the youth curfew and the City of Perth's wavering support for the Nyoongar Patrol have sparked criticism of how governments view Northbridge's diversity. Some have argued that in seeking to market the area's cultural diversity, planners have drawn on romantic notions rather than the lived realities of people on the streets. Iveson (2000:231) suggests that planners have defined diversity to include 'al fresco dining, shoppers and workers and tourists intermingling, jazz bands and farmers markets' but not Indigenous people and youth. At the same time, government efforts to improve the perceptions and realities of safety in Northbridge enhance prospects for profitability among local entrepreneurs and the enjoyment of the precinct by the majority of visitors. These are important concerns that must be weighed against the risks of ethnic stereotyping when determining a suitable approach to policing.

NORTHBRIDGE: A COSMOPOLITAN PRECINCT?

Northbridge as an ethnic precinct and residential neighbourhood has worn successive ethnic masks to match the changing composition of its residents and shop owners. Unlike the Chinatowns and Little Italys of Sydney and Melbourne, Northbridge—or more correctly, its regulators, immigrant entrepreneurs, tourist authorities, ethnic community organisations and the multicultural public who comprise Zukin's critical infrastructure—has yet to settle on its ethnic identity. The district's multi-ethnic character constrains attempts to link it to a single ethnic group. The regulators of local government (the City of Perth and the Town of Vincent) and the provincial government have played key roles in the development of Northbridge's public and private spaces. Consistent with regulation theory (Fainstein *et al.* 2003), local stakeholders (immigrant entrepreneurs, local residents and ethnic community organisations) have been able to exercise agency concerning the changing ethnic face of Northbridge.

The question of which ethnic groups are included and consulted in this re-visioning—and which are not—is politically charged. We have seen contradictions (Collins 2007) in the many incarnations of Northbridge as an ethnic precinct: its changing, multi-ethnic character that runs against a mono-ethnic place identity; conflicting political relations within ethnic communities over matters of urban development; and issues of safety within the precinct. While the major threat to safety in Northbridge is unrelated to its ethnic make-up, the youth curfew and Nyoongar Patrol suggest that Northbridge is struggling with the tension between the inclusion of all cultural groups and the maintenance of safety and comfort for Northbridge restaurant patrons.

In Northbridge, a succession of immigrant entrepreneurs (Kloosterman and Rath 2003) have played a crucial role in turning the area from a poorly regarded hinterland into a contemporary centre for tourism and cultural consumption, confirming the important yet relatively undeveloped link between ethnic diversity and tourism (Hall and Rath 2007). They have been supported by recent efforts by state and local governments to re-vision and revitalise the area, in part by drawing on its immigrant past. However, the failed attempts to create a Chinatown in Northbridge illustrate that such ethnic branding may be divisive and, when contrived, may fail to attract customers. The Chinatown project lacked legitimacy among the area's ethnic community groups and immigrant entrepreneurs.

The process of developing an ethnic identity for Northbridge is ongoing. It remains an accessible inner-city Perth neighbourhood with a reputation for good restaurants from a great diversity of cultures—a vibrant ethnic eating, shopping and entertainment strip. Its population of immigrant entrepreneurs, old established ethnic communities, newly-arrived immigrants and middle class gentrifiers creates an environment that continues to be popular with locals and tourists alike, even as the many stakeholders struggle for control and influence in determining Northbridge's future.

NOTES

1. The research for this study was funded by an Australian Research Council Linkage Grant LP0455640 'Cosmopolitan Heritage in a Multicultural Society: Ethnic Communities and the Built Environment in Australian Cities and Rural and Regional Areas'.
2. Calculations based on 2006 census data exclude non-responses.
3. Officially, Northbridge ends at Newcastle Street. However, the concentration of ethnic shops extends as far north as Brisbane Street, and many people see that as the 'real' boundary.
4. EPRA consulted with representatives from the Hellenic Association of WA, the Italian Chamber of Commerce, the Chung Wah Association and the Aboriginal Advancement Council.

REFERENCES

Anderson, K. (1990) '"Chinatown re-oriented": A critical analysis of recent redevelopment schemes in a Melbourne and Sydney enclave', *Australian Geographical Studies: Journal of the Institute of Australian Geographers*, 28(2): 137–54.

Australian Bureau of Statistics (2007) *2006 Census of Population and Housing, Basic Community Profile, Northbridge WA (State Suburb), BO9 Country of Birth by Person by Sex, Place of Usual Residence.* <http://www.abs.gov.au/websitedbs/d3310114.nsf/Home/census> (accessed 4 January 2008).

Blagg, H. and Valuri, G. (2004) 'Aboriginal community patrols in Australia: Self-policing, self-determination and security', *Policing & Society*, 14(4): 313–28.

Body-Gendrot, S. (2003) 'Cities, security, and visitors: Managing mega-events in France', in Hoffman, L., Fainstein, S. and Judd, D. (eds) *Cities and Visitors: Regulating People, Markets, and City Space*. Oxford: Blackwell, 39–52.

Burayidi, M. (2003) 'The multicultural city as planners' enigma', *Planning Theory & Practice*, 4(3): 259–73.

Chinatown (1982) *Chinatown: A Submission to the Federal Minister for Immigration and Ethnic Affairs*. Unpublished.

Choi, C.Y. (1975) *Chinese Migration and Settlement in Australia*. Sydney: University of Sydney Press.

City of Perth (2007) *Northbridge Piazza.* <http://www.perth.wa.gov.au/web/Council/Plans-and-Projects/Current-Projects/Northbridge-Piazza/> (accessed 9 January 2008).

———. (1989) *Northbridge Study: Discussion Paper Number One, Issues*. Perth: City of Perth.

Church, J. (2005) *Per L'Australia: The Story of Italian Migration*. Carlton: The Miegunyah Press.

Collins, J. (2007) 'Ethnic precincts as contradictory tourist spaces', in Rath, J. (ed.) *Tourism, Ethnic Diversity and the City*. New York: Routledge, 52–67.

———. (2005) *Ethnic Precincts as Contradictory Tourist Spaces: The Case of Sydney, Australia*. Paper at COMPAS/ISCA Seminar Series, Oxford University, 10 June.

———. (2002) 'Chinese entrepreneurs: the Chinese Diaspora in Australia', *International Journal of Entrepreneurial Behaviour & Research*, 8(1/2): 113–33.

Collins, J., Gibson, K., Alcorso, C., Tait, D. and Castles, S. (1995) *A Shop Full of Dreams: Ethnic Small Business in Australia*. Sydney and London: Pluto Press.

Collins, J. and Kunz, P. (2007) 'Ethnic entrepreneurs, ethnic precincts and tourism: The case of Sydney, Australia', in Richards, G. (ed.) *Tourism, Creativity and Development*. London and New York: Routledge, 201–14.

———. (2005) *Spatial Dimensions of the Commodification of Ethnicity in the City: Producers, Consumers and the Critical Infrastructure in Four Sydney Ethnic Precincts*. Paper at IMISCOE Cluster B6 Workshop on Ethnic, Cultural and Religious Diversity, Amsterdam, 26–28 May.

Collins, J., Nobel, G., Poynting, S., and Tabar, P. (2000) *Kebabs, Kids, Cops and Crime: Youth, Ethnicity and Crime*. Sydney: Pluto Press.

Cox, S. (2003) *North of the Line: The 'Demonisation' of Northbridge Youth and the 'Exclusion' of Aboriginal and 'Asian' Youth*. Unpublished thesis, Curtin University of Technology.

de Lepervanche, M. (1975) 'Australian immigrants 1788 to 1940: Desired and unwanted', in Wheelwright, E.L. and Buckley, K. (eds) *Essays in the Political Economy of Australian Capitalism Volume 1*. Sydney: Australia and New Zealand Book Company, 72–104.

Dunn, K. (2003) 'Using cultural geography to engage contested constructions of ethnicity and citizenship in Sydney', *Social & Cultural Geography*, 4(2): 153–65.

East Perth Redevelopment Authority (EPRA) (2000a) *The History of Northbridge*. East Perth: East Perth Redevelopment Authority.

———. (2000b) *Northbridge: The Master Plan—October 2000*. East Perth: East Perth Redevelopment Authority.

Edensor, T. (2001) 'Performing tourism, staging tourism: (Re)producing tourist space and practice', *Tourist Studies*, 1(1): 59–81.

———. (1998) *Tourists at the Taj: Performance and Meaning as Symbolic Site*. London: Routledge.

Experience Perth (2007) *Map of the City of Perth*. <http://www.experienceperth.com/en/City+of+Perth/City+Map/default.htm> (accessed 1 July 2008).

Fainstein, S., Hoffman, L. and Judd, D. (2003) 'Introduction', in Hoffman, L., Fainstein, S. and Judd, D. (eds) *Cities and Visitors: Regulating People, Markets, and City Space*. Oxford: Blackwell, 1–20.

Fitzgerald, S. (1997) *Red Tape, Gold Scissors: The Story of Sydney's Chinese*. Sydney: State Library of New South Wales Press.

Gabaccia, D. (1998) *We Are What We Eat: Ethnic Food and the Making of Americans*. Cambridge: Harvard University Press.

Government of Western Australia (2008) 'Taking people out of harms way', *InterSector*. <http://intersector.wa.gov.au/article_view.php?article_id=80&article_main=33> (accessed 21 April 2008).

Gregory, J. (2007) *Northbridge and Carlton: A Tale of Two Inner City Suburbs*. Paper at Northbridge History Studies Day, Perth, 12 May.

Hage, G. (1997) 'At home in the entrails of the West: Multiculturalism, "ethnic food" and migrant home building', in Grace, H., Hage, G., Johnson, L., Langsworth, J. and Symonds, M. (eds) *Home/World: Space, Community and Marginality in Sydney's West.* Sydney: Pluto Press.

Hall, C.M. and Rath, J. (2007) 'Tourism, migration and place advantage in the global cultural economy', in Rath, J. (ed.) *Tourism, Ethnic Diversity and the City.* New York: Routledge, 1–18.

Hoffman, L., Fainstein, S. and Judd, D. (eds) (2003) *Cities and Visitors: Regulating People, Markets, and City Space.* Oxford: Blackwell.

Ip, D. (2005) 'Contesting Chinatown: place-making and the emergence of "ethnoburbia" in Brisbane, Australia', *GeoJournal*, 64: 63–74.

Iveson, K. (2000) 'Beyond designer diversity: Planners, public space and a critical politics of difference', *Urban Policy and Research*, 18(2): 219–38.

Judd, D.R. (2003) 'Visitors and the spatial ecology of the city', in Hoffman, L., Fainstein, S. and Judd, D. (eds) *Cities and Visitors: Regulating People, Markets and City Space.* Oxford: Blackwell, 23–38.

King, S. (1998) *Restricted Entry: Investigating Chinese Immigration to Western Australia.* Perth: National Trust of Australia (WA).

Kinkead, G. (1993) *Chinatown: A Portrait of a Closed Society.* New York: Harper Perennial.

Kloosterman, R. and Rath, J. (eds) (2003) *Immigrant Entrepreneurs: Venturing Abroad in the Age of Globalization.* Oxford and New York: Berg.

———. (2001) 'Immigrant entrepreneurs in advanced economies: Mixed embeddedness further explored', *Journal of Ethnic and Migration Studies*, Kloosterman, R. and Rath, J. (eds) *Special Issue on 'Immigrant Entrepreneurship*, 27(2): 189–202.

Kringas, P. (2001) 'Post-war Greek immigration', in Jupp, J. (ed.) *The Australian People: Encyclopedia of the Nation, Its People and Their Origins.* Cambridge: Cambridge University Press, 392–95.

Lash, S. and Urry, J. (1994) *Economies of Signs and Space.* London and New Delhi: Sage.

Light, I. and Gold, S. (2000) *Ethnic Economies.* San Diego: Academic Press.

Light, I. and Rosenstein, C. (1995) *Race, Ethnicity and Entrepreneurship in Urban America.* New York: Aidine de Gruyter.

Lin, J. (1998) *Reconstructing Chinatown: Ethnic Enclave, Global Change.* Minneapolis: University of Minnesota Press.

Mac Arthur, C. (2007) *The Emperor's New Clothes: The Role of the Western Australian Press and State Government in Selling the Story of the Northbridge Curfew.* Unpublished thesis, Murdoch University.

Meethan, K. (2001) *Tourism in Global Society: Place, Culture, Consumption.* New York: Palgrave.

Multicultural and Ethnic Affairs Commission of WA (MEAC) and Commonwealth Office of Multicultural Affairs (OMA) (1989) *Diversity is Great, Mate! A Study of Community Relations in an Inner-City Area of Perth, Western Australia.* Perth: MEAC.

Northbridge History Project (2007) <http://www.northbridgehistory.wa.gov.au/> (accessed 9 January 2008).

Pang, C.L. and Rath, J. (2007) 'The force of regulation in the Land of the Free: the persistence of Chinatown, Washington DC as a symbolic ethnic enclave', in M. Ruef and M. Lounsbury (eds) *The Sociology of Entrepreneurship (Research in the Sociology of Organizations 25).* New York: Elsevier, 191–216.

Peters, N. (2007) *On the Street Where You Live: Inner City Immigrant Enterprise.* Paper at Northbridge History Studies Day, Perth, 12 May. <http://www.northbridgehistory.wa.gov.au/documents/pdf/nbhdc-0436.pdf> (accessed 2 January 2008).

Poynting, S., Noble, G., Tabar, P. and Collins, J. (2004) *Bin Laden in the Suburbs: Criminalizing the Arab Other*. Sydney: Federation Press.

Price, C. (1963) *Southern Europeans in Australia*. Canberra: Australian National University Press.

Rath, J. (ed.) (2007) *Tourism, Ethnic Diversity and the City*. London and New York: Routledge.

———. (2002) *The Commodification of Cultural Resources in a Historic Inner City: Amsterdam Chinatown from an International Comparative Perspective*. Paper at 'The Future of the Historic Inner City of Amsterdam' Conference, 5–7 September, Amsterdam.

———. (ed.) (2000) *Immigrant Business: The Economic, Political and Social Environment*. Basingstoke and New York: Macmillan/St Martin's Press.

Rayner, M. (2003) 'Northbridge curfew', *Indigenous Law Bulletin*, 5(27): 9–13.

Richards, G. (2005) 'Introduction: Cultural tourism in Europe', in Richards, G. (ed.) *Cultural Tourism in Europe, Online Edition*. Association for Tourism and Leisure Education (ATLAS), <http://www.atlas-euro.org>: 8–20 (accessed 4 January 2008).

Sandercock, L. and Kliger, B. (1998a) 'Multiculturalism and the planning system, part 1', *Australian Planner*, 35(3): 127–32.

———. (1998b) 'Multiculturalism and the planning system, part 2', *Australian Planner*, 35(4): 223–27.

Stewart, S., Hanna, B., Thompson, S., Gusheh, M., Armstrong, H. and van der Plaat, D. (2003) 'Navigating the sea of diversity: Multicultural place-making in Sydney', *International Journal of Diversity in Organisations, Communities and Nations*, 3: 239–52.

Stransky, C. (2001) 'Italians in Western Australia', in Jupp, J. (ed.) *The Australian People: Encyclopedia of the Nation, Its People and Their Origins*. Cambridge: Cambridge University Press, 492–95.

Thompson, S. (2003) 'Planning and multiculturalism: A reflection on Australian local practice', *Planning Theory & Practice*, 4(3): 275–93.

Town of Vincent (2006) *New-look William Street*, media release, 29 August. <http://www.vincent.wa.gov.au/cproot/1247/11207/William%20Street%20 Upgrade.pdf> (accessed 9 January 2008).

Vertovec, S. (2007) 'Super-diversity and its implications', *Ethnic and Racial Studies*, 30(6): 1024–054.

Waldinger, R., Aldrich, H., Ward, R. and Associates (1990) *Ethnic Entrepreneurs: Immigrant Business in Industrial Societies*. Newbury Park: Sage.

Warde, A. (1997) *Consumption, Food and Taste*. London: Sage.

Warde, A. and Martens, L. (2000) *Eating Out: Social Differentiation, Consumption and Pleasure*. Cambridge: Cambridge University Press.

The West Australian (2005) '$122,000 funding cut to Nyoongar Patrol', 25 June: 54.

Yiannakis, J. (1996) *Megisti in the Antipodes: Castellorizian Migration and Settlement to W.A. 1890–1990*. Western Australia: Hesperian Press.

Zukin, S. (1995) *The Cultures of Cities*. Cambridge: Blackwell.

8 *Risotto* and *Zighiní*?

Milano's Lazzaretto between Multiculturalism and Insecurity

Roberta Marzorati and Fabio Quassoli

INTRODUCTION

Over the past decades, immigrants have grown increasingly visible in European cities. Cities are places where *difference* concentrates and expresses itself (Castells and Borja 1997) and where the 'intersection of migrant cultures . . . has produced a plethora of differentiated, hybrid, and heterogeneous cultural geographies' (Bridge and Watson 2000). Immigrant entrepreneurship alongside the commodification of cultural diversity has in some places transformed ethnic neighbourhoods into objects of the 'tourist gaze' (Urry 1990)—places of leisure and consumption where visitors can discover *difference* domesticated and adapted for Western use (Shaw *et al.* 2004; Binnie *et al.* 2006). This process, already well-known in some North American cities, is now occurring in Europe (Rath 2005; Hall and Rath 2007).

In southern European cities, the relationship between (multi-)ethnic neighbourhoods and the leisure and tourism industry is less clear cut (Nuvolati and Marzorati 2007). Urban multi-ethnic neighbourhoods such as Raval in Barcelona (Aramburu 2004; Serra 2006; Solé *et al.* 2007), Lavapiés in Madrid (Roch 2007; Solé *et al.* 2007; Cebrián and Bodega 2002), Esquilino in Rome (Casacchia and Natale 2003; Mudu 2003), Porta Palazzo in Turin (Semi 2004a, 2004b) and Mouraria in Lisboa (Oliveras in this book) are located in city centres, host mixed foreign populations (many of whom own businesses), and attract a diverse range of urban users and tourists. In many cases, immigrant businesses have played an important role in the economic, social and cultural transformation of these neighbourhoods.[1]

The formation of ethnic neighbourhoods as places of leisure and cultural consumption requires a 'social infrastructure' capable of supporting their development into tourist attractions (Rath 2005). Here 'social infrastructure' refers to those agents who mark urban space in a cultural sense, who designate parts of the city as sites to showcase 'ethnic' activity. Such infrastructures may exist in neighbourhoods that are not necessarily places of residence, but are nevertheless important sites for the ethnic and cultural self-representation of 'communities'. Thus defined, the social infrastructure is simultaneously the outcome of an unplanned and unpredictable process of urban transformation and the essential precondition for the tourist-entrepreneurial exploitation of cultural 'resources' in a given area. The outcome of complex social processes, the establishment of 'social infrastructures' requires a series of demographic, economic and political contingencies rooted in the existence of sufficiently large and internally stratified ethnic communities.

In contrast to many northern European and North American cities, migrant 'communities' in southern European countries have generally not formed *ethnic enclaves* (Wilson and Portes 1980). Spatial concentration remains low and ethnic social networks—though important and growing—are not as central as in many cities in North America (Parella 2005; Aramburu 2004). Very few neighbourhoods have become symbols of 'community' self-representation. In Italy, we see that although immigration has been ongoing for at least two decades and certain national groups (Albanians, Moroccans, Chinese and Peruvians) have consolidated their presence, the concentration of co-nationals has seldom given way to forms of residential clustering that would justify the use of the expression 'ethnic neighbourhood'.[2]

We also need to consider the role of local authorities in promoting entrepreneurial activity and ethnic neighbourhoods as cultural resources. 'Regulation of the city for the benefit of the visitors and the tourism industry' (Fainstein *et al.* 2003: 242) includes zoning laws as well as regulations to improve accessibility, to keep urban environments clean and safe, and regulations on the use of public land and resources for neighbourhood renewal. As several case studies in this book illustrate, a positive and collaborative approach towards immigrant entrepreneurship on the part of local institutions—as well as a broader multicultural/cosmopolitan vision for the city as a whole (Sandercock 1998)—are key factors in the evolution of ethnic neighbourhoods into places of leisure and consumption, and for the integration of immigrants into urban economies. Yet in many Italian cities, the contribution of immigrant entrepreneurs goes unacknowledged by both local institutions and autochthonous residents. In public discourse, neighbourhoods with high concentrations of foreign residents and users are often associated with urban blight and depicted as unsafe. Instead of being seen as sources of diversified cultural capital, immigrant residents are frequently represented in highly stereotyped, racialised images as 'social problems' or 'threats' to public safety (Aramburu 2002; Dal Lago 1999; Petrillo 2003; Quassoli 2004; Santamaría 2002).

This article focuses on the Lazzaretto neighbourhood of Milan, which has been a multicultural area since the 1970s.[3] In it we can observe features identified in the literature (Rath 2005; Hall and Rath 2007) as key factors in the transformation of ethnic neighbourhoods into tourist attractions. This transformation, however, has not begun in Lazzaretto; nor will it in the near future. This is due to Milan's anti-immigrant atmosphere, the local government's lack of commitment to multiculturalism, constraints on immigrant entrepreneurship placed by the institutional framework and conflict between Italians and immigrants over the use of public space.

FROM 'CASBAH' TO TRENDY NEIGHBOURHOOD

Lazzaretto occupies a semi-central location between the ramparts of Porta Venezia and the Central Station, to the west of the Corso Buenos Aires.[4] The neighbourhood is compact and densely built, consisting of about 20 regular city blocks. It was built in the second half of the nineteenth century when the Italian Credit Bank bought the area and gave the go-ahead to build high-density rental houses for middle and lower-middle class residents (Marzorati 2009: 121).

Lazzaretto has been a destination for newcomers for over half a century. Migrants from the surrounding mountain and rural areas arrived in the 1950s and 60s, followed by migrants from southern Italy. Since the 1970s,

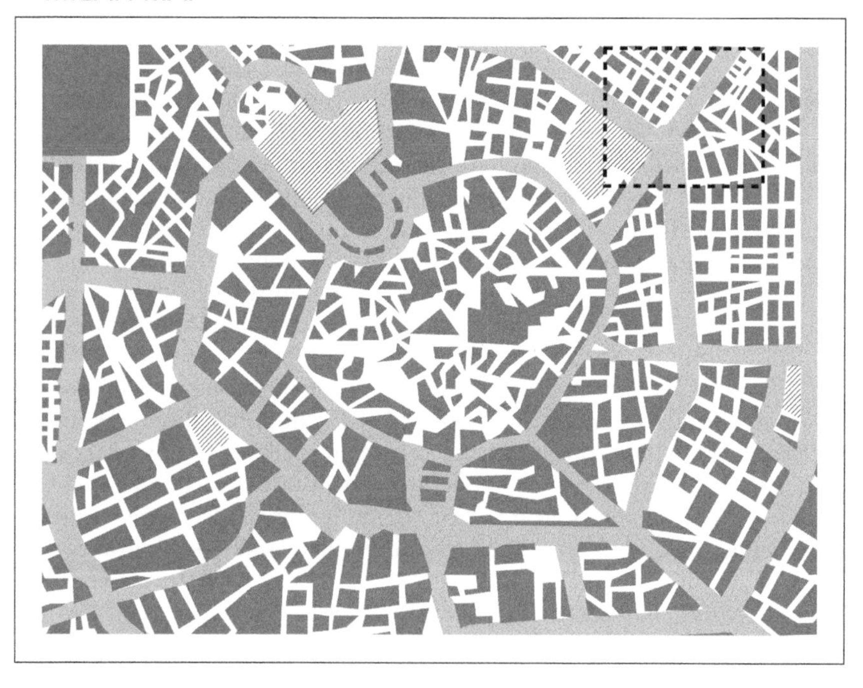

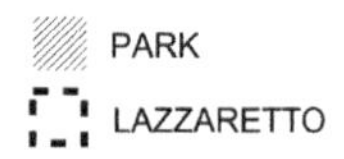

Figure 8.1 Location of Lazzaretto in Milan. *Map designed by Ana Sala.*

Lazzaretto has attracted immigrants from the former Italian colonies in Africa; with the proliferation of bars and restaurants, it has become a place of residence, consumption and social life for the Eritrean community.[5] Other groups followed the Eritreans—Algerians and Moroccans prominent among them. During this period, the local press often referred to the neighbourhood as a 'casbah', both in a folkloric sense and to highlight the presence of illicit trades and foreign as well as autochthonous crime (Cologna *et al.* 1999). More recently, other groups from Asia, Latin America and Africa have settled in the neighbourhood, leading to the burgeoning of businesses owned and patronised by foreigners.

Residentially, the area has transformed over the past decades. Until 20 years ago, densely populated and often dilapidated nineteenth-century houses accommodated both lower-middle class families from the south of

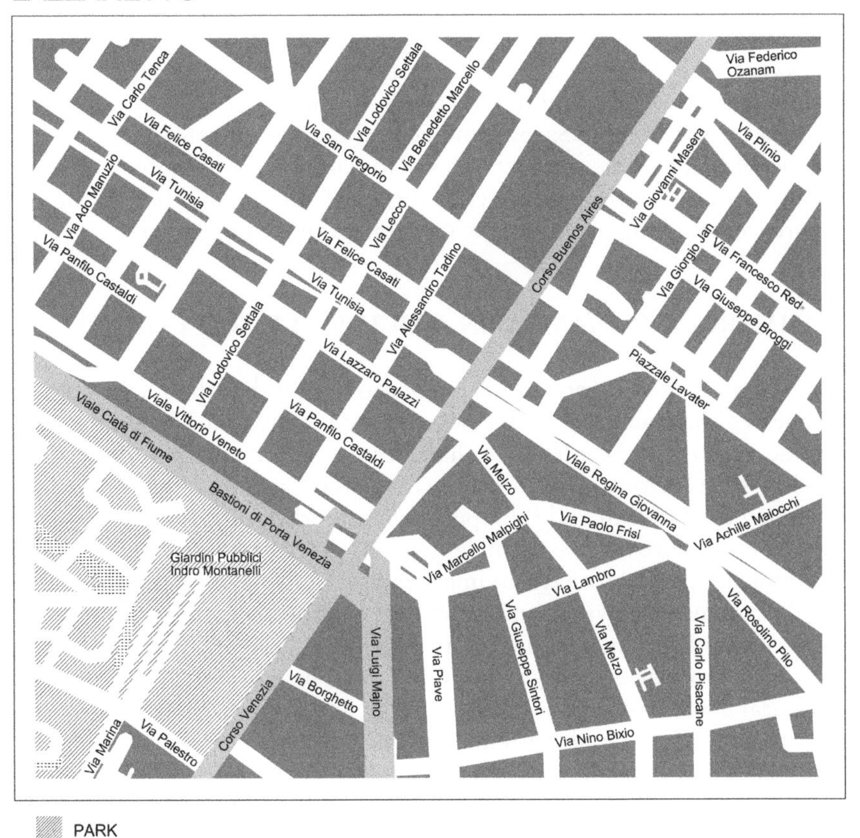

Figure 8.2 Map of Lazzaretto. *Map designed by Ana Sala.*

Italy and foreign immigrants. More recently, the restoration of residential buildings and streets 'have reshaped the neighbourhood as a whole and have triggered perceivable changing dynamics in the resident foreigners, resulting in a rising rate of well-off Italian units' (Novak 2007).

These changes can be understood against the background of more general processes of residential dispersion affecting many areas of Milan, which have been driving the lower and middle classes away from the central and semi-central areas of the city towards the suburbs. Though signs of urban segregation remain weak, a trend towards growing residential polarisation, comprised of difficult to detect micro-territorial processes, is jeopardising the traditional mixed composition of the urban fabric (Ranci 2005; Zajczyk 2003). Lazzaretto is a good example of such micro-territorial transformations as gentrification triggers conflict between old and new residents and users of public space and fuels protests by well-off Italian residents against urban 'decay' and lack of safety.

Lazzaretto is an important place for the different national groups that work and socialise there.[6] Though in ways that can hardly be compared to cities in North America and northern Europe (Shaw *et al.* 2004), Lazzaretto—due to its 'social infrastructure'—is the Milanese neighbourhood (together with the Paolo Sarpi-Canonica's Chinatown) that comes closest to the model of an ethnic precinct as a site for cultural consumption. It is thus an ideal case for examining the extent to which the exploitation of cultural differences is taking place and the constraints that can prevent it.

COMMERCIAL DYNAMISM IN A MULTICULTURAL CONTEXT

Lazzaretto is a commercially heterogeneous neighbourhood, in both its businesses and clienteles. Businesses catering mainly to foreign immigrants (phone centres, bars, groceries, hairdressers) exist alongside specialised services such as tour operators and commercial agents. Other businesses, including the Eritrean and Ethiopian restaurants that Lazzaretto is famous for,[7] cater to more diverse clienteles and are based on the commodification of ethnic qualities. More recent immigrants are also part of the neighbourhood's commercial dynamism. Of the 90 enterprises counted in 2003, 21 were run by Eritrean and Ethiopian nationals, 12 by Chinese and 12 by Indians: a growing diversity that testifies to an increasingly Asian and multicultural neighbourhood (Granata, Novak and Polizzi 2003).

Although the pace of development has slowed since the 1990s, Lazzaretto remains a dynamic place characterised by new 'activities and populations, frequent changes of business, successes and failures of frontier trades and delocalisation trends' (*Ibid.*: 127).[8] Eritreans and Ethiopians have recently resumed investing in the neighbourhood, both economically and symbolically. After years of local economic stagnation, a new generation of entrepreneurs is opening call centres, rotisseries, restaurants and bars, thus demonstrating the symbolic importance of the neighbourhood to the two 'communities'.

The entry of new entrepreneurs has strained social relations in Lazzaretto, once considered a 'neighbourhood of encounter and exchange' (Novak 2007). Established Eritrean and Ethiopian entrepreneurs have long complained about the newcomers, especially the new ethnic bar owners whose establishments attract rowdy customers. Besides disturbing the peace of the neighbourhood, these customers drive away the Italian clientele. B., who runs Lazzaretto's oldest Eritrean restaurant, describes the situation as follows:

> This neighbourhood is no longer what it once was because of the vultures . . . people who smelled the neighbourhood and decided to open clubs and stole the jobs from those doing them before . . . institutions are building a casbah to mess things up . . . politics is dirty . . . they have been well covered up by somebody powerful, by politicians because they are corrupt. (interview, Eritrean restaurant owner).

Another local entrepreneur emphasised the slump in prices caused by the new Asian businesses and the difficulties competing with them. In addition to the complaints against 'newcomers', old local entrepreneurs blame both the economic crisis and the introduction of the Euro for difficulties in their once flourishing businesses. They also blame the municipality for neglecting the neighbourhood, security and the proliferation of illegal businesses.

A more complete picture of local commerce requires mention of the increasing number of businesses run by Italians. Trendy, exotic restaurants and nightclubs as well as ethnic crafts shops run by Italians have recently opened alongside the businesses owned by foreigners, proving that Lazzaretto has gained a certain stature and that entrepreneurs are beginning to grasp its potential. Lazzaretto's commercial renewal has benefited from the 'drawing power of Corso Buenos Aires' as well as from a number of urban renewal initiatives and the vitality of ethnic commerce (Granata, Novak and Polizzi 2003). Many new commercial outlets have sprung up next to the crafts shops, the taverns and the Milanese restaurants, and are trying to exploit the economic potential of the neighbourhood. D., the owner of a trendy nightclub, imagines the future of the neighbourhood:

> I hope they will open other places like mine . . . the point is that if there are many exclusive places, even small ones, they upgrade the neighbourhood, they bring people in, and this helps everyone, the activities work better . . . but still not like the Navigli,[9] because on the weekend they are so trashy . . . I hope it will be possible to upgrade the neighbourhood, I really do hope so . . . even thought some interesting realities already exist here . . . in Lecco Street they opened a new big furniture store, and a beautiful architect's studio . . . these types of

stores last for three or four months . . . they don't work [referring to a small Latin American store that closed down, A.N.] (interview, bar owner and resident).

ETHNIC ENTREPRENEURSHIP IN LOCAL AND NATIONAL CONTEXT

Sound policies to promote tourism, leisure and cultural consumption in multi-ethnic/multicultural neighbourhoods do not exist at the city level; the foreign entrepreneurs we interviewed were generally unhappy with the city council's efforts to support businesses in Lazzaretto. This lack of action by the municipal authorities is surprising given the growing importance of immigrant entrepreneurs in Milan, whose dynamism over the last decade has contributed significantly to the city's economy. According to a recent survey of the Milanese province conducted by Formaper (an agency of the Milan Chamber of Commerce), on 31 December 2006 there were 18,992 businesses owned by immigrants from low-income countries, accounting for 7.1 per cent of all local enterprises and 46 per cent of all immigrant-owned businesses in Lombardia. About half of all immigrant-owned businesses were concentrated in a few sectors: building, street trade, transport and cleaning services. The food industry also played an important role, with about 700 immigrant-owned restaurants and 500 bars (one out of three restaurants in Milan is 'ethnic'). Immigrant groups with the most entrepreneurs included the Egyptians, Moroccans, Chinese, Romanians and Albanians, whose activities are concentrated mainly in Milan (Rosso and Soru 2007).

Taking heed of the growing role of immigrant entrepreneurs in the urban economy, the Milan Chamber of Commerce recently authorised an innovative experiment. Jointly established by public and private partners,[10] the Association for the Development of Immigrant Business in Milan (ASIIM) supports immigrant-owned businesses through training courses and an information desk aimed at 'supporting immigrants [in] creating and developing their own business'. ASIIM has instituted a council of immigrant entrepreneurs to publicise its services through ethnic networks.[11]

Together with the National Tourism Research Institute (Isnart) and the Milan Provincial Commercial Business Association (Epam), the Milan Chamber of Commerce recently published the guidebook *Milano Multietnica*. In advertising the ethnic products and services available in the city, the guidebook awards a series of restaurants the Quality Seal of the National Tourism Research Institute. This, the promoters claim, shows that 'the rich cultural diversities of the city are able to arouse tourists and citizens' interest and curiosity'.[12] These boards promoting Milan's ethno-cultural businesses comprise the core of the city's 'critical infrastructure'—agents able to influence the popularity of particular cultural products and images of areas as places for 'ethnic' consumption (Zukin 1991, 1995).

Tourist guidebooks also play a role in promoting Lazzaretto, albeit a limited one. While most guides (foreign language ones in particular) ignore Lazzaretto and Milan's other ethnic neighbourhoods, these areas are occasionally recommended to visitors wishing to get off the beaten track:

> African, Argentinean and Chinese restaurants, together with a wide range of ethnic shops, offer a varied and picturesque urban-scape further enhanced by the presence of the large street market in Via Benedetto Marcello (Istituto Geografico De Agostini 2005: 162, author's translation).

> During the 1980s, the Ethiopian-Eritrean community settled in the surroundings of Porta Venezia, close to Corso Buenos Aires. Particularly within the orthogonal grid between via Vitruvio and viale Vittorio Veneto, shops and restaurants offering Horn of Africa cuisine have opened. In the course of the following years, other foreigners—mainly Arabs and Indians—have arrived in the area, which remains the historic seat of Milan's first ethnic restaurants (Touring Club Italiano 2006: 42, authors' translation).[13]

Apart from these activities promoting immigrant-owned businesses to tourists, the political commitment is variable. On the one hand, the Chamber of Commerce, the Milan Provincial Council and the Lombardy Regional Council seem to appreciate the opportunities for urban development that immigrant-owned businesses provide, and respond to requests coming from both Italian and foreign entrepreneurs. On the other hand, the city council is notable in its absence; during interviews, members of ASIIM were hard pressed to name any city council measures that supported immigrant businesses. At the time of interviewing (November 2007), our respondents were still waiting for the newly elected city councillor for Work Policies and Occupation to name a council member to participate on the ASIIM board. The municipal administration seems to believe that supporting the development of ethnic or multicultural precincts is neither feasible nor advisable; their arguments are clearly 'couched in revanchist discourses and geared to assimilating and even disciplining minority groups' (Hall and Rath 2007: 17). The councillor responsible for tourism, identity and marketing in Milan spoke as follows:

> I don't think that immigrant entrepreneurial activities will become a local marketing tool. . . . I can't see the Sarpi Canonica neighbourhood as Milan's Chinatown, I mean Paolo Sarpi and Canonica are Milanese neighbourhoods and I don't think that the concentration of homogeneous groups, especially in a single area, would increase the neighbourhood's attractiveness . . . or, rather, maybe it could become more attractive for tourists, but it would inconvenience lawful local residents. . . . Someone who is accustomed to the corner baker, to the

> dry cleaner and the Tuscany restaurant down the street, and who the following year finds three kebab houses, won't enjoy it as a new experience, he will rather think it is a radical change in his neighbourhood, in his habits. If he wanted to buy his bread roll, he would have to walk for three kilometres, because in the kebab shop there is no choice apart from kebab food . . . and, above all, the thing that in my opinion harms multi-ethnicity, meant as a resource, the most is that there are too many illegal immigrants in Milan, too many people that . . . damage other ethnic communities already living there (interview, councillor).

If we consider the economic incorporation of immigrants in Italy through the 'mixed embeddedness' approach (Kloosterman and Rath 2003; Rath 2002)—which examines immigrants' embeddedness within social networks as well as within wider political and economic structures—the peculiarities of the Milanese case become easier to understand. On the one hand, immigrant businesses have not attained the critical threshold necessary for the creation of real ethnic economies in the tourism, leisure and culture sector. On the other hand, there is a local political system that—far from seizing the economic and cultural potential of these changes—fans widespread fears regarding the multicultural transformation of urban areas.

Our study also confirms one of the main features of immigrants' socio-economic integration in Italy over the past two decades. The demand for labour in manufacturing has provided immigrants, particularly in Italy's northern regions, with numerous opportunities to enter the labour market as unskilled workers without provoking conflict between them and their domestic colleagues (Ambrosini 1999; Magatti and Quassoli 2003; Reyneri 1998). Until recently, immigrants did not have access to self-employment and entrepreneurship. Italy is characterised by a much higher percentage of self-employed workers and small businesses than other developed countries; in many Italian regions—certainly in the richest ones—it has been an overt strategy of the middle and lower-middle classes to start their own businesses in order to gain access to otherwise unattainable income and status levels (Magatti and Quassoli 2003). The relatively high prestige associated with being a small entrepreneur in many regions of the centre and north of Italy means that the view of the immigrant as an employee rather than as a self-employed person is widespread. While immigration is easily seen as a source of labour in areas that would otherwise be affected by shortages, it is much more difficult to accept the idea that immigrants may start their own businesses, as Italians do. The idea of subordination is expressed chiefly by equating immigrants with employees.

Though immigrant or 'ethnic' entrepreneurship has existed in Italy for a number of years, the reciprocity clause allowing immigrants to start their own businesses in Italy only if the same right were granted to Italians in their home countries was revoked only in 1998, with the enforcement of law number 40 (known as Turco-Napolitano). While the new legislation enabled the spread of entrepreneurship and self-employment among immigrants (Chiesi 2003), permits of stay (known as *permesso di soggiorno*) to engage in self-employment are issued subject to legal and practical constraints, producing a situation that Ambrosini (2004) calls a 'normative context of partial liberalisation'. The overall situation therefore consists of steadily growing immigrant entrepreneurship, but within a social, normative and political framework that limits the development of differentiated and articulated forms of immigrant self-employment, thus confining most immigrants' initiatives to marginal sectors (Martinelli 2003).

ETHNIC NEIGHBOURHOODS: PERCEPTIONS OF INSECURITY AND NEGATIVE STEREOTYPING

The dominant public discourse associates the concentration of immigrants in certain neighbourhoods with issues of safety, urban decay and the risk of petty crime. The image of the foreigner that predominates in political and media discourse remains enmeshed in discriminatory and racist stereotyping (Maneri 2001; Faso 2008). Racism also informs standard descriptions of Lazzaretto by many shopkeepers and residents:

> Here in the evening you have only them around . . . this bar makes me sick. . . . Sometimes when you go for a walk, they are all standing around on the sidewalks. We are obliged to get off the sidewalk because they don't move an inch! And if you say 'excuse me' they don't even listen to you, and if you say something they look askance at you. So better being standoffish than getting knifed (interview, building caretaker).

Local political authorities speak in the same tone:

> This is a degraded area . . . once you get in, you never know how you will come out . . . do you remember when they occupied that house? This is an 'at risk area'. We know that we have troubles with it . . . there has been a lot of damage these days . . . if you go there on Sunday morning you'll find a sea of broken bottles (interview, Councillor of Council Area Number 3).

Research on the relationship between safety and urban transformation has repeatedly shown that the widespread feeling of insecurity in many Italian cities (particularly in the centre and north) due to the assumed increase in delinquent behaviour should in fact be associated with the types of changes often identified as the very preconditions for the emergence of ethnic neighbourhoods and their subsequent exploitation as places of leisure and cultural consumption (Petrillo 2000, 2003; Quassoli 2004). Rapid turnover in the social and demographic composition of neighbourhoods, transformation of the city's economic and commercial fabric and conflict between groups who use public spaces differently generate feelings of loss of control over the conditions in which everyday life takes place. Urban conflict often involves those who occupy a relatively advantaged position (usually those who have been residents of a neighbourhood for a longer time) invoking the intervention of public authorities to restore a social order that cannot be attained through endogenous and informal social processes. Municipal authorities are generally requested to enact repressive and preventive measures to soothe conflict and restore a semblance of control over the urban environment, 'neutralising' through police intervention those actors (graffiti writers, night revellers, migrants, etc.) who are otherwise held, through their behaviour, to undermine the safety of all (IReR 2006).

In this context, political alliances between the original residents, who tend to organise in committees to promote safety and defend their quality of life, and local authorities—particularly when they are right-wing—are the major obstacle to the development in Italy of those processes that have enhanced the exploitation of multicultural urban areas in other countries (Rath 2005; Hall and Rath 2007; Shaw *et al.* 2004). Finally, it should be added that the transformations that characterise neighbourhoods with high immigrant concentrations are difficult to control or channel through policy and urban planning, as the instruments of the latter are often too cumbersome compared to the rapidity of changes brought about by immigration. As La Cecla (1998) argues, Italy still largely associates immigrant neighbourhoods with *no-go areas* and prefers to manifest its *proteophiliac* (Bauman 1993) dispositions—to consume ethnicity and diversity protected by distance, within the safety of other, neutral parts of the city.
Many of our interviews clearly confirmed such a picture:

> I won't know what my culture is until I don't know who I am, I see all the other cultures as invaders, therefore I see them as dominants, I feel under threat . . . the Eritrean restaurant that is opening in the former tavern, where I used to eat risotto with saffron, it is threatening my identity because it uproots something that is ingrained in the city, brings something that overwhelms me. It shouldn't be like this. . . . Whichever feature is representative of Milan, it must be promoted to the utmost, in order to coexist peacefully, even with cultural contamination, bearing in mind that Milan and New York are among the non-capital cities with the greatest number of consular presences in the world, 98 to be precise, therefore it is a multi-ethnic city even in its highest rungs, we don't need to wait for regular immigration, the population here is multi-ethnic from the outset, because it welcomes a multitude of different ethnic groups, I think there are 178, our main concern is that immigration waves often raise public security-related issues (interview, councillor, Municipality of Milan).

Paradoxically, the Eritreans—who began Lazzaretto's multicultural transformation in the 1970s—are the immigrant group with the worst reputation. Due to the worsening conflict with Ethiopia and an increasingly authoritarian government, many young Eritreans have sought asylum in Italy over the past years. They face many obstacles including government red tape, social barriers and neglect of their needs by the local authorities. They thus gravitate to Lazzaretto to seek support from their compatriots, meeting in ethnic bars and on street corners; in 2005 a group of young Eritreans occupied a building. Local residents and Italian traders see them as a threat to the neighbourhood's security and, above all, to their businesses:

> I have the impression that there is absolutely no racial integration . . . and when there is no integration, it is as an end to itself, but a bad end, I see all these coloured people wandering aimlessly all day long, nobody knows how they manage to live, or what they do . . . I've been living here for seven years and I don't believe this neighbourhood has many attractive features [commercial and ethnic activities, A.N.]. I mean, to me, from the point of view of someone who's living here, and who doesn't agree with . . . then the tourist who comes from afar says: 'let's go to that multi-ethnic neighbourhood' (interview, bar owner and resident).

> It seems rather a mono-ethnic neighbourhood, than a multi-ethnic one, because there are a lot of Eritreans . . . this is not a problem for me, but for someone else it could be (interview, Argentinean restaurant employee).

We need to stress that complaints by residents and entrepreneurs do not address the immigrant presence *tout court*. What they underline—a point also confirmed by immigrant entrepreneurs—is the lack of effective regulation concerning zoning, environmental protection, the use of public spaces and control of illegal commercial activities. They also ask for greater commitment to urban renewal, considered fundamental to the neighbourhood's economic development, on the part of the public authorities.[14] All of these aspects moreover constitute some of the preconditions for the development of ethnic neighbourhoods into places of leisure and consumption (Rath 2005; Hall and Rath 2007).

CONCLUSION

A relish for so-called 'ethnic' goods has arisen in Italy as it has in the rest of the Western world. This trend expresses itself through forms of experiential consumption (Pine and Gilmore 1999) that imply the purchase not merely of goods, but of *difference* that has been domesticated and adapted for Western use (Binnie *et al.* 2006). With the 'aesthetization of everyday life' (Featherstone 1991), 'ethnic' food, clothing, furnishings and music have become objects of consumption, often associated with an elite clientele proud of its cosmopolitan and multicultural tastes. The appreciation of 'ethnic' products has not only influenced people's consumption patterns, but has contributed to the transformation of urban spaces—although this has not happened in Italy in the spectacular or ostentatious forms that encompass entire neighbourhoods (think of Chinatowns) in the United States, Canada or Australia (see for example Anderson 1990, 1995; Collins 2007; Pang and Rath 2007).

Our analysis of Lazzaretto's transformation has shown that immigrant entrepreneurs are contributing to the 'symbolic economy' of the city (Zukin 1995). Nevertheless, the complexity of the interacting processes affecting the neighbourhood make it difficult to predict how the situation will evolve.

There remain significant constraints on immigrant entrepreneurs as well as on immigrants living in Italy more generally. As in other southern European countries, some of these constraints are probably related to immigrant entrepreneurship being in its early days. However, the cultural premises concerning migrant participation in the economy and the anti-immigrant atmosphere that pervades Milanese as well as Italian political discourse cannot be discounted. Well-off Italian residents, shopkeepers and established immigrant entrepreneurs alike demand greater commitment from public institutions in regulating change. Their demands, however, do not entail the promotion of Lazzaretto as an ethnic precinct; the neighbourhood's multicultural atmosphere is often considered more of a problem than a resource as it attracts groups who negatively affect the area's image. Their claims are mainly rooted in control and safety issues, which are of course central to the development of urban neighbourhoods (ethnic or not) as places of leisure, consumption and tourism. But in Lazzaretto's case, residents and entrepreneurs are more interested in retaining their local clienteles than in welcoming tourists. For most of them, 'making the neighbourhood safer' (Quassoli 2004) means removing 'undesirable' people—the homeless, drug-addicts and, above all, foreign immigrants—from the urban landscape.

Local authorities moreover seem unengaged with changes in the neighbourhood and remain committed to a parochial and ethnocentric vision of the neighbourhood and the city. If we further consider that immigrant entrepreneurs in Lazzaretto have failed to advance a strategic vision which harnesses the commodification and promotion of cultural diversity to the area's development, the resulting local framework is very different from what has been reported in other European and North American cities. In this context, it seems highly unlikely that a local 'growth coalition' (Rath 2005; Hall and Rath 2007) will develop.

We finally return to a basic question. When we examine the literature on the commodification of cultural diversity, we come across urban and national contexts in which immigration itself is uncontroversial. Hence the analyses take for granted that both fundamental and citizenship rights are guaranteed to foreign residents, that migrants can be the legitimate targets of specific labour, social and cultural policies, and that they are a resource (in both economic and cultural terms) for society at large. This does not (yet) seem to be the case in Italy. Before betting on some of Milan's neighbourhoods becoming places of leisure and consumption through the exploitation of cultural diversity, public authorities had better focus on implementing immigrants' citizenship rights, on managing cohabitation between immigrants and autochthonous residents (which to date has been problematic), and on improving the widespread negative attitudes towards immigration.

ACKNOWLEDGMENTS

We would like to thank Jan Rath, Aytar Volkan and Eduardo Barberis for their critical comments and insightful suggestions to the article.

NOTES

1. While the presence of consumers, leisure seekers and tourists has recently grown in some southern European neighbourhoods, the reasons are not always clear. In Barcelona, urban renewal and the birth of the Cultural Quarter through the building of the Museu d'Art Contemporani de Barcelona and the Centre de Cultura Contemporània de Barcelona have made Raval into a tourist attraction (Degen 2003; García and Claver 2003). But it is unclear to what extent tourists are also attracted by the ethnic shops and restaurants that give Raval a multicultural atmosphere. The case of Porta Palazzo in Turin seems quite different as parts of the neighbourhood have been gentrified and ethnicity is part of the discourse concerning the requalification of the area (Semi 2004a, 2004b).
2. The Chinatowns established by Chinese immigrants in Milan, Rome, Prato and Naples are among the rare exceptions to this pattern. Senegalese immigrants have also tended to congregate in certain neighbourhoods, in some cases even in single buildings, with cases registered in the Provinces of Bergamo, Brescia, Ravenna and Rimini (Sinatti 2006).
3. Empirical evidence for our case study was collected during 2007. Most of the interviews were conducted in November and December 2007.
4. Corso Buenos Aires is one of Milan's main shopping streets. See <http://ciaomilano.it/e/shops/baires.asp>.
5 Eritreans are among the oldest immigrant groups in Italy. A first immigration wave began in the 1960s following annexation by Ethiopia, when single or small groups of Eritrean women went to Italy to work for Italian families. From 1975 until the end of the 1980s, civil war fuelled a mass exodus. A well-organised refugee network was established across Europe and political parties/movements involved in the war against Ethiopia (the Eritrean Liberation Front and the Eritrean People's Liberation Front) became active in immigration countries. Eritreans living in Milan used to hand over part of their revenues to the EPLF, while Eritrean restaurants in Lazzaretto were controlled by EPLF representatives. An economic immigration wave began after the war ended in 1991, but as the political situation in Eritrea worsened, it

was again fuelled by political factors. Despite the historical relations between Italy and Eritrea, Eritreans living in Italy enjoy no special legal status.

6. Besides the streets (where Eritreans and Ethiopians usually meet), the Church of San Carlino al Lazzaretto is a major meeting point for Sinhalese, as is the Orthodox Church for Russians, Ukrainians and Moldavians.
7. During the second half of the 1990s, about 15 of the 40 foreign businesses in Lazzaretto and beyond Corso Buenos Aires were East African restaurants.
8. The 'Indian Shop' is a typical case: for 26 years it was a neighbourhood landmark, though it has since been passed down to Pakistanis, Indians and Bengalis.
9. 'Navigli' is probably the city's busiest nightlife area.
10. Chamber of Commerce, Bocconi University, Milan Town Council, Milan Provincial Council, Lombardia Region, Assolombarda, Commerce, Tourism, Services and Professions Union of Milan Provincial Council, Craft Union of Milan Provincial Council, and Bank of Milan.
11. There are further initiatives by immigrant entrepreneurs, including *Impresa Etnica*, an online newspaper that aims to be 'a technical and professional information tool to give voice and to foster the integration of both women and immigrant entrepreneurs'. While such initiatives appear positive, particularly in light of the widespread public mistrust of immigrant entrepreneurs, they show that the phenomenon of immigrant-owned businesses in Milan and Italy is in its initial stage.
12. <http://www.mi.camcom.it/show.jsp?page=703030>.
13. Although Milan is firmly on the tourist circuit—in 2004, 11.8 million foreign tourists visited Rome, 5.6 million visited Venice, 4.7 million visited Florence and about 4 million came to Milan (Marra 2007)—most visitors are 'business tourists'. Milan continues to emphasise its role as Italy's economic capital; shopping for 'made in Italy' fashion icons and designer products is a main attraction, implying well-heeled visitors. There is also a shortage of affordable accommodation, rendering Milan out of reach to many young visitors (those most likely to be attracted by a multicultural offer). Policies to develop previously neglected sectors have focused on art (e.g. Da Vinci's Last Supper) rather than on ethnic promotion. Lazzaretto must also compete with well-established nightlife areas such as I Navigli, l'Isola and Corso Como: historic locations that offer a wide choice of bars, restaurants and nightclubs (see Bovone 1999).
14. In a *Corriere della Sera* article titled 'Lazzaretto becomes a ghetto', an Italian shopkeeper proposes making via Lecco a pedestrian street to make the area more attractive. In the same article, a local politician proposes fiscal incentives to facilitate the arrival of 'new and sane commercial activities' (see D'amico 2008).

REFERENCES

Ambrosini, M. (2004) *Gli immigrati nelle attività indipendenti: interpretazioni a confronto, Crocevia Working Paper.* <http://www.fieri.it/download.php?fileID=147&lang=ita> (accessed 27 May 2009)

———. (1999) *Utili invasori. L'inserimento degli immigrati nel mercato del lavoro italiano*. Milano: Franco Angeli.

Anderson, K.J. (1995) *Vancouver's Chinatown: Racial Discourse in Canada, 1875–1980*. Montreal: McGill-Queens University Press.

———. (1990) 'Chinatown re-oriented: A critical analysis of recent redevelopment schemes in a Melbourne and Sydney enclave', *Australian Geographical Studies*, 28(2): 137–154.

Aramburu, M. (2004) 'Los comercios de inmigrantes extranjeros en Barcelona y la recomposición del inmigrante como categoría socia', *Scripta Nova. Revista Electrónica de Geografía y Ciencias Sociales*, 6(108). <http://www.ub.es/geocrit/sn/sn-108.htm> (accessed 27 May 2009).

———. (2002) *Los 'Otros' y 'nosotros'. Imágenes del inmigrante en Ciutat Vella de Barcelona*. Madrid: Ministerio de Educación y Cultura.

Bauman, Z. (1993) *Postmodern Ethics*. Oxford and Cambridge: Blackwell.

Binnie, J., Holloway, J., Millington, S. and Young, C. (2006) *Cosmopolitan Urbanism*. London and New York: Routledge.

Bovone, L. (ed.) (1999) *Un quartiere alla moda. Immagini e racconti del Ticinese a Milano*. Milano: Franco Angeli.

Bridge, G. and Watson, S. (2000) 'City differences', in Bridge, G. and Watson, S. (eds) *A Companion to the City*. Oxford: Blackwell, 251–60.

Camera di Commercio di Milano (2007) *Milano Multietnica*. Milano: Actl.

Casacchia, O. and Natale, L. (2003) 'L'insediamento degli extracomunitari a Roma: un'analisi sul rione Esquilino', in Morelli, R., Sonnino, E. and Travaglino, C.M. (eds) *I territori di Roma*. Rome: Università di Roma, 610–39.

Castells, M. and Borja, J. (1997) *Local & Global*. Madrid: Taurus.

Cebrián, J.A. and Bodega, M.I. (2002) 'El negocio étnico, nueva fórmula de comercio en el casco antiguo de Madrid: el caso de Lavapiés', *Estudios Geográficos*, 63(248/249): 559–80.

Chiesi, A.M. (2003) 'Problemi di rilevazione empirica del capitale sociale', in Andreotti, A. and Barbieri, P. (eds) *Reti e capitale sociale, Inchiesta*, 139 (special issue): 86–97.

Collins, J. (2007) 'Ethnic precincts as contradictory tourist spaces', in Rath, J. (ed.) *Tourism, Ethnic Diversity and the City*. New York: Routledge, 67–86.

Cologna, D., Breveglieri, L., Granata, E. and Novak, C. (eds) (1999) *Africa a Milano. Famiglie, ambienti e lavori delle popolazioni africane a Milano*. Milano: Abitare Segesta Cataloghi.

Dal Lago, A. (1999) *Non Persone: l'esclusione dei migranti in una società globale*. Milano: Feltrinelli.

D'amico, P. (2008) 'Negozi murati e bivacchi nei giardini Così l'ex Lazzaretto diventa ghetto', *Corriere della Sera*, 4(May): 9.

Degen, M. (2003) 'Fighting for the global catwalk: Formalizing public life in Castlefield (Manchester) and diluting public life in el Raval (Barcelona)', *International Journal of Urban and Regional Research*, 27(4): 867–80.

Fainstein, S., Hoffman, L. and Judd, D. (2003) 'Making theoretical sense of tourism', in Fainstein, S., Hoffman, L. and Judd, D. (eds) *Cities and Visitors: Regulating People, Markets and City Space*. Oxford and Cambridge: Blackwell, 239–253.

Faso, G. (2008) *Razzismo democratico. Le parole che escludono*. Rome: Derive e Approdi.

Featherstone, M. (1991) *Consumer Culture and Postmodernism*. London: Sage.

García, M. and Claver, N. (2003) 'Barcelona: Governing coalitions, visitors and the changing city center', in Fainstein, S., Hoffman, L. and Judd, D. (eds) *Cities and Visitors: Regulating People, Markets and City Space*. Oxford and Cambridge: Blackwell, 113–25.

Granata, E., Novak, C. and Polizzi, E. (2003) 'Immigrazione dall'Asia e trasformazione urbana', in Cologna, D. (ed.) *Asia a Milano: famiglie, ambienti e lavori delle popolazioni asiatiche a Milano*. Milano: Abitare Segesta Cataloghi, 96–151.

Hall, M. and Rath, J. (2007) 'Tourism, migration and place advantage in the global cultural economy', in Rath, J. (ed) *Tourism, Ethnic Diversity and the City*. New York: Routledge, 1–24.

Istituto Geografico de Agostini (2005) *Milano*. Novara: De Agostini.

IReR (2006) *Costruzione di un sistema per l'analisi dei rischi e la formulazione di scenari nell'ambito della sicurezza urbana*. Unpublished research report <http://www.irer.it/ricerche/istituzionale/programmazione/2005B035> (accessed 27 May 2009).

Kloosterman, R. and Rath, J. (eds) (2003) *Immigrant Entrepreneurs: Venturing Abroad in the Age of Globalization*. Oxford and New York: Berg.

La Cecla, F. (1998) 'L'urbanistica è di aiuto alle città multietniche?', *Urbanistica*, 111: 45–47.

Magatti, M. and Quassoli, F. (2003) 'Italy: Between legal barriers and informal arrangements', in Kloosterman, R. and Rath, R. (eds) *Immigrant Entrepreneurs: Venturing Abroad in the Age of Globalization*. London: Routledge, 147–172.

Maneri, M. (2001) 'Il panico morale come dispositivo di trasformazione dell'insicurezza', *Rassegna Italiana di Sociologia*, 1(1): 5–40.

Marra, E. (2007) *Marketing Urbano: comprendere le nuove potenzialità di attrazione della città*. Unpublished paper.

Martinelli, A. (2003) 'Imprenditorialità etnica e società multiculturale', in Chiesi, A.M. and Zucchetti, E. (eds) *Immigrati imprenditori. Il contributo degli extracomunitari allo sviluppo della piccola impresa in Lombardia*. Milano: Egea.

Marzorati, R. (2009) *Lontani Vicini. processi di costruzione sociale dell'alterità in contesti locali: una comparazione fra Milano e Barcellona*. Unpublished PhD thesis, Università degli Studi di Milano Bicocca and Universitat Autònoma de Barcelona.

Mudu, P. (2003) 'Gli Esquilini: contributo al dibattito sulle trasformazioni nel rione Esquilino di Roma degli anni Settanta al Duemila', in Morelli, R., Sonnino, E. and Travaglino, C.M. (eds) *I territori di Roma*. Rome: Università di Roma, 641–80.

Novak, C. (2007) 'Abitare in un quartiere multietnico', in Multiplicity Lab (eds) *Cronache dell'abitare Milano*. Milano: Bruno Mondadori, 221–31.

Nuvolati, G. and Marzorati, R. (2007) 'Quartieri etnici fra conflitti e city marketing', *Sociologia Urbana e Rurale*, 83: 61–84.

Pang, C.L. and Rath, J. (2007) 'The force of regulation in the Land of the Free: the persistence of Chinatown, Washington D.C. as a symbolic ethnic enclave', in M. Lounsbury and M. Ruef (eds) *The Sociology of Entrepreneurship* (Research in the Sociology of Organizations, Vol. 25). New York: Elsevier, 191–216.

Parella, S. (2005) 'Estrategias de los comercios étnicos en Barcelona, España', *Política y Cultura*, 23: 257–75.

Petrillo, A. (2003) La città delle paure: per un'archeologia delle insicurezze urbane. Avellino: Sellino Editore.

———. (2000) *La città perduta: l'eclissi della dimensione urbana nel mondo contemporaneo*. Bari: Edizioni Dedalo.

Pine, B.J. II and Gilmore, J.H. (1999) *The Experience Economy: Work is Theatre & Every Business a Stage*. Boston: Harvard Business School Press.

Quassoli, F. (2004) 'Making the neighbourhood safer: Social alarm, police practices and immigrant exclusion', *Journal of Ethnic and Migration Studies*, 30(2): 1163–181.

Ranci, C. (2005) 'Problemi di coesione sociale a Milano', in Magatti, M., Ceruti, M., Senn, L., Balducci, A., Artoni, R., Sapelli, G., Ranci, C., Manghi, B., Dente, B., Colombo, A., and Ciborna, C. (eds) *Milano nodo della rete globale: un itinerario di analisi e di proposte*. Milano: Bruno Mondadori.

Rath, J. (2005) 'Feeding the festive city: Immigrant entrepreneurs and tourist industry', in Guild, E. and van Selm, J. (eds) *International Migration and Security: Opportunities and Challenges*. London and New York: Routledge, 238–56.
———. (ed.) (2002) *Unravelling the Rag Trade: Immigrant Entrepreneurship in Seven World Cities*. Oxford and New York: Berg.
Reyneri, E. (1998) 'The role of the underground economy in irregular migration to Italy: cause or effect?', *Journal of Ethnic and Migration Studies*, 24(2): 313–331.
Roch, F. (2007) 'La ciudad histórica como lugar para la convivencia. Inmigración y vida urbana en el barrio de Lavapiés de Madrid', in Grandi, F. and Tanzi, E. (eds) *La città meticcia: riflessioni teoriche e analisi di alcuni casi europei per il governo locale delle migrazioni*. Milano: Franco Angeli.
Rosso, A. and Soru, A. (2007) *L'imprenditorialità degli immigrati in Lombardia—Indagine quantitativa*. Progetto Equal Ministero del Lavoro e Regione Lombardia. <http://www.formaper.it/index.phtml?Id_VMenu=470> (accessed 27 May 2009).
Sandercock, L. (1998) *Towards Cosmopolis: Planning for Multicultural Cities*. London: Wiley.
Santamaría E. (2002) *La incognita del extraño: una aproximación a la significación sociológica de la 'inmigración no comunitaria'*. Barcelona: Anthropos.
Semi, G. (2004a) *Il multiculturalismo quotidiano: Porta Palazzo tra commercio e conflitto*. Unpublished PhD dissertation, Università di Torino and EHESS-Paris.
———. (2004b) 'Il quartiere che (si) distingue. Un caso di "gentrification" a Torino', *Studi Culturali*, 1(1): 83–107.
Serra, P. (2006) *El comercio étnico en el distrito de Ciutat Vella de Barcelona*. Barcelona: Fundació 'la Caixa'.
Shaw, S., Bagwell, S. and Karmowska, J. (2004) 'Ethnoscapes as spectacle: Reimaging multicultural districts as new destinations for leisure and tourism consumption', *Urban Studies*, 41(10): 1983–2000.
Sinatti, G. (2006). 'Diasporic cosmopolitanism and conservative translocalism: Narratives of nation among Senegalese migrants in Italy', *Studies in Ethnicity and Nationalism*, 6(3): 30–50.
Solé, C., Parella, S. and Cavalcanti, L. (2007) *El empresariado inmigrante en España*. Barcelona: Fundación 'La Caixa'.
Touring Club Italiano (2006) *Milano*. Milano: TCI.
Urry, J. (1990) *The Tourist Gaze: Leisure and Travel in Contemporary Societies*. London: Sage.
Wilson, K. and Portes, A. (1980) 'Immigrant enclaves: An analysis of the labor market experiences of Cubans in Miami', *American Journal of Sociology*, 86(2): 295–319.
Zajczyk, F. (2003) 'Segregazione spaziale e condizione abitativa a Milano', in Negri, N. and Saraceno, C. (eds) *Povertá e vulnerabilitá sociale in aree sviluppate*. Rome: Carocci, 55–75.
Zukin S. (1995) *The Cultures of Cities*. Cambridge and Oxford: Blackwell.
———. (1991) *Landscapes of Power: From Detroit to Disney World*. Berkeley: University of California Press.

Contributors

Volkan Aytar, previously at the Turkish Economic and Social Studies Foundation (TESEV), Bahçeşehir University and Sabancı University (all in Istanbul) and the State University of New York at Binghamton, is now with the Institute for Migration and Ethnic Studies (IMES) at the University of Amsterdam and Bahceşehir University in Istanbul. An Editorial Board Member of *Istanbul Journal of Urban Culture*, his interests include forms of employment in Istanbul's entertainment and tourism sectors, and the social construction of spaces of music and consumption and their symbolic meanings. His book *Metropol* (Metropolis) was published by L&M Publishing House in 2006. His articles have appeared in, among others, *Geocarrefour: Revue de Geographie de Lyon*, *Dialoghi Internazionali: Citta nel Mundo* and *Itinerari d'Impresa*. He has contributed to edited volumes by Immanuel Wallerstein and Jan Rath.

Sue Bagwell is the Research Development Manager for the Cities Institute (http://www.citiesinstitute.org) at London Metropolitan University, where she is currently engaged in a number of research studies for local regeneration agencies and research councils. Her research interests include black and ethnic minority enterprises (BEMEs) and enterprise support policies and their evaluation. She has undertaken research and development work with Vietnamese businesses in both Vietnam and the UK and is currently involved in research on key business clusters in the City Fringe area of London as part of a four-year evaluation of the City Growth Strategy for the area.

Jock Collins is Professor of Economics at the University of Technology (UTS) in Sydney. His research interests centre on the interdisciplinary study of immigration and cultural diversity in the economy and society. His recent research has been on Australian immigration, ethnic crime, immigrant entrepreneurship, ethnic precincts and tourism, and the social use of ethnic heritage and the built environment. He has published extensively in the field, with eight books, numerous articles in national and international journals, and chapters in books. His last book, co-authored with others, is *Bin Laden in the Suburbs: Criminalising the Arabic 'Other'* (Federation Press 2004).

Kirrily Jordan was with the School of Finance and Economics at the University of Technology in Sydney and is now a Postdoctoral Fellow at The Australian National University's Centre for Aboriginal Economic Policy Research. She is co-author of a recent book on economic inequality in Australia and has published regularly on inequality, ethnicity and urban environments.

Süheyla Kırca-Schroeder is Associate Professor of Media and Cultural Studies at Bahcesehir University, Istanbul, and is currently a Visiting Professor in the European Media Studies Department at Potsdam University. Her research interests include popular culture and media, representations of gender and ethnicity, the globalisation of culture, transnational Turkish media, media integration policies in Germany and the construction of social identities in different cultural spheres. She has published extensively in the field of media and cultural studies. In addition to her articles in national and international journals, Kirca Schroeder is the author of *Popüler Feminizm: Britanya ve Türkiye'de Kadın Dergileri* (Popular Feminism: Women's Magazines in Britain and Turkey, 2007), and the co-editor of *Dışarıda Kalanlar/Bırakılanlar* (Outsiders, 2001), *Freedom and Prejudice: Approaches to Media and Culture* (2008) and *Beyond Boundaries: European Media, Culture and Identity* (2009).

Roberta Marzorati is Postdoctoral fellow at the Department of Sociology of the University of Milano-Bicocca. She got her PhD in Urban and Local European Studies from the same university and in Sociology from the Universitat Aurónoma de Barcelona where she worked with the Grup d'Estudis d'Immigració i Minories Ètniques (GEDIME). Her main research interests are urban immigration, multicultural neighborhoods (especially in southern European cities) and the social construction of Otherness.

Johannes Novy received his PhD in Urban Planning from Columbia University's Graduate School for Architecture, Planning and Preservation in 2011 and currently works as a researcher at the Center for Metropolitan Studies in Berlin. He is the co-editor of "Searching for the Just City" (Routledge 2009); enjoys working on and writing about urban (development) politics in the United States and Europe; the role of tourism and leisure in processes of contemporary urban change, as well as issues relating to urban social justice and activism.

Catarina Reis de Oliveira is Head of the Office of Research and International Relations at the *Alto Comissariado para a Imigraçã e Diáogo Intercultural* (ACIDI) in Portugal. Until 2005 she was a Lecturer in Sociology at the New University of Lisbon. In 2004–2008 she was involved in co-ordinating the EU Ethnic Minority Entrepreneurs Network. She has published extensively on immigrant entrepreneurship, including book chapters, journal articles and two books: *Estratégias Empresariais de Imigrantes em Portugal* (2004) and *Empresários de Origem Imigrante* (2005). In 1999–2000 she was awarded a Merit Scholarship by the New University of Lisbon and an academic award in Multiculturalism and Ethnicity in Contemporary Society by the Gulbenkian Foundation.

Ching Lin Pang was postdoctoral researcher at Migration and Ethnicity Research Institute Brussels (MERIB), Catholic University of Leuven, Belgium, and is now at the Center for Equal Chances and against Racism, Brussels. She has previously researched Japanese and Chinese immigrants in Belgium, with a particular emphasis on their ethnic identity and social position in both receiving and host societies. She has also recently started to study (female) immigrant entrepreneurship and new migrations. She is the author of *Negotiating Identity in Contemporary Japan: The Case of Kikukoshijo* (2000) and co-editor of *Structure, Culture and Beyond* (1999) and *Cultuur, Etniciteit en Migratie* (1999). She was a member of the TSER Thematic Network 'Working on the Fringes'.

Fabio Quassoli is Professor of Sociology at the University of Milano-Bicocca. His research interests centre on the sociology of immigration, multicultural societies and inter-cultural communications. He is currently working on immigrant policies and the social construction of immigrant crime in Italy. He has published three books, numerous articles in national and international journals and chapters in books. His last book is *Riconoscersi. Differenze culturali e pratiche comunicative* (Raffaello Cortina Editore, 2006).

Jan Rath is Professor of Urban Sociology, Head of the Department of Sociology and Anthropology, and Researcher at the Institute for Migration and Ethnic Studies (IMES) at the University of Amsterdam (*http://www.janrath.com*). An anthropologist and urban studies specialist, he is the author, editor or co-editor of numerous articles, book chapters, reports and books on the sociology, politics and economics of post-migration processes. They include *Immigrant Businesses: The Economic, Political and Social Environment* (Macmillan, 2000); *Unravelling the Rag Trade: Immigrant Entrepreneurship in Seven World Cities* (Berg/New York University Press, 2002); *Immigrant Entrepreneurs: Venturing Abroad in the Age of Globalization* (Berg, 2003), *Tourism, Ethnic Diversity, and the City* (Routledge, 2007), *Ethnic Amsterdam* (Amsterdam University Press, 2009), and *Selected Studies in International Migration and Immigrant Incorporation* (Amsterdam University Press, 2010).

Stephen Shaw is Director of the TRaC Research Centre, and Senior Lecturer at the Cities Institute, London Metropolitan University. His research topics include the development of sustainable visitor economies in disadvantaged areas, the built heritage of immigrant communities, public spaces and inter-culturalism. In his previous career, he worked as a Chartered Town Planner, and has worked closely with practitioners to engage urban communities in schemes to make the public realm more accessible, safe and welcoming. He is a member of the Cultural Tourism Committee of ICOMOS UK (UNESCO World Heritage), and chairs the Canada-UK Cities Research Group.

Index

For Product Safety Concerns and Information please contact our EU representative GPSR@taylorandfrancis.com Taylor & Francis Verlag GmbH, Kaufingerstraße 24, 80331 München, Germany

T - #0121 - 160425 - C0 - 229/152/10 - PB - 9780415719681 - Gloss Lamination